SpOOky
SCIENCE

SPOOKY SCIENCE

DEBUNKING THE PSEUDOSCIENCE OF THE AFTERLIFE

JOHN GRANT

STERLING
New York

STERLING
New York

An Imprint of Sterling Publishing
1166 Avenue of the Americas
New York, NY 10036

ISBN 978-1-4549-1654-3

Distributed in Canada by Sterling Publishing
c/o Canadian Manda Group, 664 Annette Street
Toronto, Ontario, Canada M6S 2C8
Distributed in the United Kingdom by GMC Distribution Services
Castle Place, 166 High Street, Lewes, East Sussex, England BN7 1XU
Distributed in Australia by Capricorn Link (Australia) Pty. Ltd.
P.O. Box 704, Windsor, NSW 2756, Australia

For information about custom editions, special sales, and premium
and corporate purchases, please contact Sterling Special Sales at
800-805-5489 or specialsales@sterlingpublishing.com.

BOOK DESIGN BY ANNA CHRISTIAN

Manufactured in the United States of America

2 4 6 8 10 9 7 5 3 1

www.sterlingpublishing.com

Art on facing page: A 1922 "spirit" photograph with a (living) Sir Arthur Conan
Doyle in the foreground. Doyle fell hook, line, and sinker for spirit photography,
as he did for the famous photographs of the Cottingley fairies (page 58). *Art on
next page:* An 1899 photograph purportedly showing a ghost. The effect was
achieved through double exposure.

For Cameron and Ditz

CONTENTS

INTRODUCTION

"It is not unlikely that more than ninety percent of paranormal literature is rubbish."

—Sir Kelvin Spenser, Foreword to *From Enigma to Science* (1973) by George W. Meek

THE HISTORY OF SCIENTIFIC ATTEMPTS TO RESEARCH the paranormal can be considered to have had two primary phases. The first began in 1882 with the formation of the Society for Psychical Research (SPR), which came about as a response to the craze for Spiritualism that had been going on ever since the Hydesville rappings case of 1848 (page 2). While originally concerned with attempts to communicate with the dead, testing spirit mediums for fraudulence, and the like, the SPR eventually turned its attention to matters like telepathy and clairvoyance.

The second phase began in 1927, when the American parapsychologist J. B. Rhine set up his laboratory at Duke University in North Carolina. His focus was on bringing scientific experimentation to bear on those purported faculties of mind that the SPR had begun to explore. Effectively, that second phase is still underway, although the work of Rhine himself is held in far lower regard today than it once was. His methods of testing by card-guessing have largely fallen into disuse, but his notion that the paranormal should be investigated by science in the same way as any other phenomenon still holds sway.

Evolutionary biologist Stephen Jay Gould famously posited that science and religion were two "non-overlapping magisteria"—a fancy way of saying that both are different ways of explaining the world, and neither has very much of value to say about the other. A similar stance could be taken in the context of science and the supernatural. One operates according to rules of reason and logic; the other does not. Throughout much of human history, however, people have assumed that the supernatural is as real as tangible reality, and have tried to use the tools of rationalism to understand and tame it. On occasion these efforts have been successful, such as when people have discovered that something they thought was part of the supernatural's magisterium—like comets or, for some TV presenters, the tides—was in fact perfectly explicable in terms of straightforward physics.

A prime example of a phenomenon that was once thought to be supernatural being demystified through the application of the scientific method is mesmerism. The tale of how this came about is by no means a simple one of science's bright light banishing the shadows.

☛ MESMERISM

Born in 1734 in Konstanz, in what's now Germany, Franz Anton Mesmer graduated with a degree in medicine from the University of Vienna. In his degree thesis he proposed that human health could be affected by the gravitational influences of the celestial bodies through the medium of the "ether," a sort of undetectable fluid that Isaac Newton had suggested might fill the universe and facilitate transmission of light, heat, gravity, and magnetism. Around this time it was thought that magnets might have healing properties, and Mesmer saw how this seemed to dovetail with his own ideas about the effects of gravity on health; gravity and magnetism were, he reasoned, really just different forms of each other. He thus set about trying to cure people by rubbing strong magnets over them. (This practice still survives in the "alternative" technique known as magnetotherapy.)

In due course, primarily because others were complaining that he'd stolen their ideas, Mesmer concluded that actual magnets weren't required. Human beings possessed a hitherto unrecognized *spiritual* form of magnetism that he called "animal magnetism" (from *anima*, the soul). His new therapy involved waving his hands

A contemporary wood engraving by H. Thiriat showing Franz Anton Mesmer and his assistant at work with a group of aristocratic patients.

around close to, but not actually touching, the patient's body. Like any quack, Mesmer reported sensational results. But in reality some of his results were good, and these occurred when his patients fell into a trance.

In 1785 the French established two commissions to examine the phenomenon of animal magnetism, the more important of the two being that under the auspices of the Académie des Sciences, which numbered Antoine Lavoisier and Benjamin Franklin among its members. It concluded forthrightly that there was no such thing as animal magnetism, and that any cures Mesmer was achieving were effected by the patients' own minds. (Nowadays we might attribute this to the placebo effect.) The same went for the mesmeric trances into which some patients fell. The commission's conclusions came agonizingly close to identifying what was really going on, but backed off just before hitting upon the modern concept of *suggestion*.

In the United Kingdom, the idea of animal magnetism never really caught on; its light flickered briefly but was extinguished by the end of the eighteenth century. However, in the 1830s, John Elliotson—professor of medicine at University College, London, and a man who was much engaged by another pseudoscience whose light briefly shone, phrenology[1]—became enamored of animal magnetism. When the French mesmerist Jules Denis, Baron du Potet, visited London, Elliotson and he conducted experiments in animal magnetism in the public wards of University College Hospital. Since it would be hard to find more suggestible people than the impoverished occupants of nineteenth-century public wards, and since everyone could see how everyone else was reacting to the activities of the two physicians, it's not surprising that these experiments were regarded by Elliotson as a resounding success, with lots of trances and (at least temporary) cures. It wasn't long before he was putting on public demonstrations . . . at which point the University College authorities declared themselves no longer amused and fired him.

A few years later, while argument still raged over the existence or nonexistence of Mesmer's soul magnetism, the Scottish surgeon James Braid decided to investigate for himself. He discovered that the mesmeric trance could be induced by insisting that the subject stare intently at something in a way that required effort of the eye muscles—an object above the natural line of sight or the famous swinging/twirling pocket watch, for example. He assumed the trance was a consequence of the exhaustion produced by this exertion, and called the state "nervous sleep." And so was mesmerism tamed from the wild realm of the paranormal by the agency of scientific reason.

Except that the story doesn't stop there. Like Mesmer before him, Braid began to assume he could use hypnotism to treat just

---

[1] Phrenology is the idea that a person's character and more could be assessed by measuring the shapes and sizes of the bumps on his or her skull. Although this might seem silly to us now, in the context of medical knowledge at the time it was far less so.

A FULL DISCOVERY

OF THE

STRANGE PRACTICES

OF

Dr.ELLIOTSON

On the bodies of his

FEMALE PATIENTS!

AT HIS HOUSE, IN CONDUIT STREET, HANOVER SQ·

WITH ALL THE SECRET

EXPERIMENTS HE MAKES UPON THEM,

AND THE

Curious Postures they are put into while sitting or standing, when awake or asleep!

A female Patient being blindfolded, to undergo an operation.

THE WHOLE AS SEEN

BY AN EYE-WITNESS,

AND NOW FULLY DIVULGED!

&c. &c. &c.

Title page of a sensationalist anonymous tract "exposing" supposed sauciness in John Elliotson's medical treatments.

about any condition with which he was presented: epilepsy, anky-losing spondylitis, congenital deafness, and the like. There seemed to be almost nothing impervious to Braid's new therapy. Also, he was as wildly enthusiastic a supporter of phrenology as Elliotson had been before him. He would put a subject into a trance, touch a bump on the subject's skull, and convince himself that the subse-quent behavior or attitude of the entranced subject matched the emotion that was indicated by the standard phrenological interpre-tation of that bump.

Because of Braid's preoccupation with trying to marry hypnosis and phrenology, it was left to others—notably Thomas Laycock of Edinburgh University and William Benjamin Carpenter of University College, London—to recognize and expand upon the breakthrough that Braid had made. Both men realized that what hypnosis offered was a temporary suspension of the will's control over the subject's actions. This allowed direct access to—and opened up for study—what Carpenter called "unconscious cerebration" and what Laycock referred to, in terms that correspond more closely to modern ones, as the brain's "reflex functions."

What do we mean by a reflex func-tion of the brain? A lot of automatic actions—breathing, maintaining the heartbeat, digestion, and so on—are under the control of the spinal cord (autonomic nervous system). But many of the actions that we regard as auto-matic are far more complex than this and require the intervention of the brain. For example, if I decide to sit down in a chair, this seems like a simple enough task. For my brain, however, it involves the elaborate and coordinated

William Benjamin Carpenter, who helped bring hypnosis into the realm of science.

adjustment and rearrangement of untold numbers of muscles, according to biochemical changes coded by carefully designed neurological signals—a process of extraordinary complexity that my brain performs without any conscious instruction from me.[2] The difference between this sort of automatic behavior and, say, breathing is that I do have to contribute that first conscious decision to sit down; when it comes to breathing, not only does it happen without my conscious intervention, but it actually requires an effort of will to stop doing it. In a hypnotic trance, my own will has effectively left the premises, so that all the automatic functions of my brain are at the disposal of the hypnotist's will.

The distinguished neurophysiologist Marshall Hall led the resistance to the new ideas of Laycock and Carpenter, and here we find ourselves back in the midst of the battle between the supernatural and the scientific. Hall, a devout Christian, was concerned that science was making it more and more difficult to retain belief in the existence of the human soul. He was therefore determined to mark off the brain as the soul's territory—the elaborate nexus from which the soul sprang. But, if the brain was capable of supporting the soaring soul, surely it was inconceivable that it could be doing something so prosaic as controlling the act of sitting down. Hall's response was completely irrational, but people who find their beliefs in the intangible threatened by science and reason very often do react irrationally.

The process of dragging the status of hypnotism from paranormal to pseudoscientific to scientific has been a tangled endeavor. It was, however, possible; and it was brought about by gradual advances

---

[2] One of the most complicated tasks for the designers of humanoid robots is to give their creations the ability to climb a flight of stairs. This requires immensely complicated instructions from the robot's central computer. Yet climbing stairs is something most of us do without devoting much thought to it.

in our understanding. The breakthrough came with Braid's discovery that the phenomenon was genuine. By contrast, in other areas of the supernatural—and indeed the paranormal in general—investigators have yet to reach that very first, essential stage: the clear demonstration that there is a phenomenon to investigate. It's difficult not to conclude, after so much effort over such a long period, that they may never do so.

✦ SO WHY DO WE FALL FOR IT?

Whenever a series of previously vaunted psychic investigations (or paranormal experiments) is shown to be erroneous or fraudulent, even though most professional researchers quite correctly abandon any claims they might have made based on those reports, there's still usually a core group that clings to the belief that somehow the results can still be believed: "Just because so-and-so cheated sometimes doesn't mean he cheated *all* the time." "Who's to know that the debunking itself isn't flawed or that the debunkers don't have an agenda of their own?" It typically takes a good long while for people committed to the field, whether professionally or merely as interested observers, to accept that they've been gulled. Just look at those who clung to their belief in the psychic abilities of the Fox sisters, even after the sisters themselves had publicly admitted their trickery (page 6). Even most researchers who do readily admit that long-treasured experiments, like the mid-twentieth-century researches of S. G. Soal into ESP, were produced fraudulently and must be abandoned, and that there are hundreds and thousands of similar examples out there, will simply start citing as support someone else's experiments that have yet to be exposed as bogus. This is a wonderful defense mechanism for psychic research: Unless and until every single relevant report has been revealed as a farrago, the mantra runs, there's no reason to believe that the supernatural is bunk.

Imagine if any scientific discipline followed these same rules. When, for example, it was discovered in the mid-1990s that the nineteenth-century biologist Ernst Haeckel had doctored much of the evidence for his biogenetic law ("ontogeny recapitulates phylogeny"), the only people trying to salvage Haeckel's reputation were (paradoxically) the Creationists, who needed him as a straw man so they could attack evolutionary biology. Biologists didn't look around in an attempt to find a replacement Haeckel; besides, they'd rejected the biogenetic law long ago on the grounds that it failed to withstand a stack of contrary evidence. Somehow ideas like telepathy and clairvoyance—not to mention life after death—seem never to be held to the same standard. Fraud? Adverse evidence? Hey, but look over here: *This* guy hasn't been debunked yet.

We tend to think that suggestibility is something that other people are susceptible to, but in reality we're all vulnerable to it. In his book *Paranormality* (2011), psychologist Richard Wiseman recalls showing a group of volunteers a glass bottle containing green liquid

The discredited biologist Ernst Haeckel, 1896. Why isn't psychic research held to the same standards as orthodox science?

and telling them it was essence of peppermint. He was going to open the bottle and wanted people to raise their hands when the peppermint fumes reached them. Sure enough, even though the green liquid was, in fact, odorless, hands started going up—first in the front row, next in the row behind. He also recounts a more ambitious version of this experiment done in the 1970s by University of California scientist Michael O'Mahoney, with the collaboration of the BBC. In a live broadcast,

O'Mahoney displayed a Rube Goldberg–like device that he said was a prototype of a gadget capable of transmitting odors through people's TV sets and into their living rooms. Soon people started phoning in to say that, yes, the experiment had been a success. Indeed, suggestion (in the psychological sense) is a tool that phony psychics deploy artfully and often with very great success.

In my impressionable youth, I managed to persuade myself that "paranormal faculties" like ESP and precognition could have a rationalist basis. It occurred to me that perhaps both these abilities could be related to pattern recognition: that the successful "telepath" or "clairvoyant" or "precognitor" was simply recognizing patterns that most of us didn't notice. For example, given enough (near infinite!) information, the future becomes predictable. By this logic, if you're skilled enough, consciously or (more likely) unconsciously, at recognizing the patterns of the present, you should be able to make a pretty good guess about the future.

In everyday life, our ability to recognize patterns quickly and thus figure out cause and effect is of great use to us, and may even be essential to our survival. We know that plunging a stick into a wasp's nest is not a good idea because, even if we haven't seen for ourselves the subsequent pattern of events, other people have described it to us—many of whom themselves perhaps may not have witnessed the consequences in person. Even when our pattern-recognizing ability gets it wrong, most often this is not a bad thing. The classic scenario is that of one of our forebears out on the veldt when he heard the grasses rustling nearby. There were two likely causes of that rustling: the wind, or a hungry saber-toothed cat. Our forebear, if he had any sense, assumed the pattern of events—grasses, the rustling—reflected the presence of the cat, and backed off: Better to flee from the wind than not to flee from the cat. But sometimes we see patterns that simply aren't there. As with the wind and the saber-toothed cat, most of the time this false recognition doesn't matter much, but sometimes it can severely mislead us—and even lead us into catastrophic danger. Here the example is of the guy who falls off the top of the Empire State Building and, halfway down, can't understand what all the fuss is about . . . because the pattern of events so far is that he's doing just fine.

We tend to try to interpret everything that happens to us as a narrative: We make stories out of our lives, and if necessary we distort or change the facts until they fit our narratives. I've called this process by the (ugly) name *narrativization*, and we'll come across it from time to time. Narrativization is one of the reasons why eyewitness evidence is so unreliable, despite the reverence in which it's held by courts of law.

Stage magicians exploit our tendency to recognize patterns and thereby wrongly attribute causes and effects: "The conjurer never saw the card I picked and therefore couldn't have known what it was." The fraudulent psychic capitalizes on this same effect to rip us off. We might think that misinterpreting a pattern of circumstances to make people think they've seen ghosts or UFOs is unimportant, but, taken together, all these false reports add up to a huge and threatening mudslide that swamps the attention of a population that might better direct its attentions to more crucial matters, such as the survival of future generations. There's only so much brainrot a society can take before it faces the threat of collapse.

RATTLING THE BONES

SPIRITUALISM AND THE REALM OF THE DEAD

"That part—talking to the dead—is easy. It's the getting-a-response part that's a little tricky."

—Richard Roeper, *Debunked!* (2008)

"People Who Don't Know They're Dead: How They Attach Themselves to Unsuspecting Bystanders, and What to Do about It."

—Title of book by Gary Leon Hill (2005)

I N 2003, A SURVEY DONE BY THE BARNA RESEARCH Group of Ventura, California, showed that 81 percent of Americans believe in life after death. Leaving aside the "don't know's," just 10 percent responded that they thought death was The End. By contrast, a 1978 Gallup survey showed that two-thirds of Americans believed in survival after death, with one-fifth dismissing the concept. In other words, belief that we live on after death appears not to be decreasing, as we might expect, but actually to be increasing. There may be all sorts of factors at work here, but part of the change is almost certainly that we seem to be returning to what astronomer Carl Sagan so memorably called the "demon-haunted world." Those TV shows promoting phony mediums are definitely having an effect.

❧ RAPPINGS

Hydesville, New York, 1848. The home of the Fox family appears to have become haunted, and the family's two younger girls—Kate and Margaret (Maggie), ages twelve and fifteen—are communicating with the spirits of the dead using a system of coded knocks. The neighborhood is agog with the tale, and in due course the story of the

Hydesville rappings becomes a sensation throughout much of the world. The echoes still resonate.

Despite popular belief, the Hydesville rappings did not mark the start of the Spiritualist craze; the girls had had plenty of precursors on both sides of the Atlantic. Among these was the English religious sect called the "Shaking Quakers," or Shakers. The Shaker "Mother Ann" Lee emigrated to America in 1774 and made a name for herself as a medium, just in time for the religious outbreak of 1779. Although she died in 1784, plenty of others followed in her footsteps. From about the 1830s, it became customary at Shaker services and even just in daily life for people to find the spirits talking through them, either coherently or in tongues. Long before the Foxes, then, there was a not-insignificant sect in the United States that had spirit medium-ship as a focal point of its beliefs. Even so, it was the names of the Fox girls that were on everyone's lips.

Within two decades of the Hydesville cause célèbre, some 30 percent of the US population believed that communication with the dead had been achieved. If you wanted to communicate with your own dear departed, that wouldn't pose much of a hardship because, out of a population of some thirty-five million, no fewer than sixty thousand were spirit mediums—that is, about one person out of every six hundred.

Celebrated journalist Horace Greeley was certain of the veracity of the Fox sisters. In *Modern American Spiritualism* (1869), Emma Hardinge quotes him thus: "Whatever may be the origin or cause of the 'rappings,' the ladies in whose presence they occur do not make them. We tested this thoroughly and to our entire satisfaction. Their conduct and bearing is as unlike that of deceivers as possible . . . And it is not possible that such a juggle should have been so long perpet-uated in public." So much was he impressed that he invited Kate to stay with his family the following fall.

At the time, Kate was conducting some of the many séances she did for wealthy New York widower Charles F. Livermore, whose

MISS MARGARETTA FOX.   MISS CATHARINE FOX.   MRS FISH.

PUB. BY N. CURRIER.   *Entered according to Act of Congress in the year 1852 by N. Currier in the Clerk's Office of the District Court of the Southern District of N.Y.*   152 NASSAU ST. COR. OF SPRUCE

MRS FISH AND THE MISSES FOX,

THE ORIGINAL MEDIUMS OF THE MYSTERIOUS NOISES AT ROCHESTER WESTERN, N.Y.

beloved wife Estelle had died in the late 1850s. In late 1871, in the 388th of these séances, Estelle announced that this was to be her last appearance. By now Kate's alcoholism had become so egregious that the kindly Livermore sent her off on a cruise to the United Kingdom in hopes that this would bring her out of herself. It was thus that she came to meet the celebrated physical chemist Sir William Crookes, who was at the time an active figure in the SPR (page 20).

Earlier the Fox sisters had subjected themselves and their rappings to investigation by a committee of three doctors from the University of Buffalo's School of Medicine. The committee concluded that the Foxes' performances were bogus. But the committee's explanation for how the trick was done—given by one of them, Austin Flint, in the *Quarterly Journal of Psychological Medicine and Medical Jurisprudence* for July 1869—was itself almost as bogus:

> *The displacement occasioning the knockings is sufficient to remove the ridge of bone which divides the two articular surfaces of the upper extremities of the tibia, from its situation in the sulcus between the condyles of the femur, and to carry it, more or less, upon the surface of the outer condyle. This movement gives rise to the first sound, and the return of the bone to its place causes its second sound, which, in the Rochester knockings, generally follows quickly on the first. We are unable to explain fully the precise mechanism by which the displacement is effected . . .*

Fighting our way through the anatomical terminology, we discover that what Flint is trying to say is that the rappings were effected by Maggie Fox dislocating her kneecap and then allowing it to snap back into place. As Flint confessed, "That sounds so loud should originate

A Currier & Ives print of Margaret (left) and Kate (center) Fox, with their married sister Leah Fish, 1852.

in the way we have ascertained that they are produced, would surprise even the medical listener, and perhaps seem almost incredible."

When, decades later, Maggie Fox confessed in a statement in the October 21, 1888, issue of the *New York World* how the trick had in fact been done, the explanation bore little resemblance to Flint's:

> *At night, when we went to bed, we used to tie an apple on a string and move the string up and down, causing the apple to bump on the floor, or we would drop the apple on the floor, making a strange noise every time it would rebound. My mother listened to this for a time. She could not understand it and did not suspect us of being capable of a trick because we were so young. . . . At last she could stand it no longer and she called the neighbors in and told them about it. . . .*
>
> *My sister Kate was the first one to observe that by swishing her fingers she could produce certain noises with the knuckles and joints, and that the same effect could be made with the toes. Finding that we could make raps with our feet—first with one foot and then with both—we practiced until we could do this easily when the room was dark. No one suspected us of any trick because we were such young children.*

Of course, there were—and are—plenty in the Spiritualist fraternity who claim that Maggie's confession was itself a falsehood perpetrated because, by now alcoholic and destitute, she was eager to lay her hands on the money the *New York World* offered her. There's some truth to this, as the money was almost certainly what prompted Maggie to confess all to the newspaper, but the confession itself seems quite genuine. She added that:

> *Spiritualism is a fraud of the worst description. I have had a life of sorrow, I have been poor and ill, but I consider it my*

duty, a sacred thing, a holy mission to expose it. I want to see the day when it is entirely done away with. After my sister Katie and I expose it, I hope Spiritualism will be given a death blow.

One of the countless popular songs inspired by the Spiritualist movement, "Spirit Rappings" (1853) by J. Ellwood Garrett (The publisher got his initial wrong on the cover.) and W.W. Rossington.

Yet, incredibly, that wasn't quite the end of their careers as mediums. Established in 1884 under the auspices of the University of Pennsylvania—which had received funding for the project to the tune of $60,000 (about $1.5 million today) in the will of the industrialist Henry Seybert—the Seybert Commission set out with the purported aim of investigating the most plausible current supernatural claims. As its first report revealed all too clearly, the commission was of a skeptical bent. Among its experimental subjects was Maggie Fox, despite her and Kate's prior confession. When the Seybert investigators tested her, they were treated initially to a succession of loud knocks that she insisted were the spirit of Henry Seybert trying to communicate his desire that the commission investigate her impartially. The men then asked her to stand on a platform of glass tumblers, which made it a little more difficult to produce floor-shaking knocks. Or did it? She professed herself able to hear further knocks, even though no one else in the room could. Unsurprisingly, the commission dismissed her claims.

Maggie and Kate cooperated with Reuben Briggs Davenport when he wrote his short book *The Death-Blow to Spiritualism* (1888), in which he detailed many of the tricks the girls had used and exposed the credulity of those around them, from their family and neighbors in the early days to such eminent figures as Sir William Crookes later on. He waxes lyrical in his opening pages:

> *That the inventors of an infamous fraud should deal to it its death-blow, is the poetic justice of fate.*
>
> *Over the creature, the creator has power of life and death.*
> *The creators of Spiritualism abjure its infamy.*
> *They decree its death.*
> *They condemn it to final destruction.*
> *They fasten upon those who continue to practice it the obloquy of history, and the scorn of mankind for all time to come.*

Margaret and Catherine Fox, the youngest of three sisters, were the first to produce "spiritualistic manifestations."

They are now the most earnest in denunciation of those impostures; the most eager to dissipate the foolish belief of thousands in the flimsiest system of deception that was ever cloaked with the hypocrisy of so-called religion.

When, as by accident, they discovered a method of deceiving those around them by means of mysterious noises, they were but little children, innocent of the thought of wrong, ignorant of the world and the world's guile, and imagining only that what they did was a clever lark, such as the adult age easily pardons to exuberant and sprightly youth.

Not to them did the base suggestion come that this singular, this simple discovery, should be the means of deluding the world, of exalting them in the minds of the weakly credulous and of bringing them fame and splendor and sumptuous pleasure.

No one who learns their true history can still believe them guilty of the willful inception of this most grotesque, most transparent and corrupting of superstitions.

The idea had its monstrous birth in older heads, heads that were seconded by hearts lacking the very essence of truth and the fountain of honest human sympathy.

❧ SCIENTISTS AND SPIRITUALISM

Although it's repeated several times in this book, there's a point that's worth making again here: When they stray outside their own areas of expertise, scientists are little better qualified to comment on matters than are lay observers—at least, those lay observers who possess the skill of logical reasoning—and they may actually be worse off, in that intelligent amateurs are less likely to fall into traps set by

⇉ MICHAEL FARADAY ⇇

An early debunker of Spiritualism was the great English physicist Michael Faraday. In 1853 he published a piece in a London literary magazine, the *Athenæum*, describing his investigations of the table-tilting phenomenon, whereby a table moved under the fingertips of a circle of sitters in response to interrogation. Faraday constructed various devices to test for supernatural agencies, such as a special table whose top consisted of two horizontal boards with rollers in between. This setup was sensitive to the slightest of involuntary movements made by his experimental subjects, who sat in a circle around it, their hands on the upper surface.

Sure enough, even though the laboratory "sitters" could have sworn they had stayed perfectly still, the device recorded plenty of movements: Involuntary muscle flexings and the like were responsible for these movements, not the activities of entities from the great beyond. Add to the mix, in the ordinary séance, the desire of at least some of the participants to achieve a positive result, not to mention possible fraud, and it isn't hard to see where table-tilting came from.

Of course, the Spiritualist faithful responded angrily to Faraday's assertions, along the usual lines: Just because tables *could* be influenced this way didn't mean the spirits weren't doing it, too! Faraday's response was one that recurs again and again in skeptical considerations of the psychic: In general, the parsimonious explanation—the one that's economical in that it relies least upon unexplained factors or a requirement that the laws of science be rewritten—is almost always the correct one.

intellectual overconfidence. And, as practitioners from Harry Houdini to Harry Gordon to Milbourne Christopher to James Randi and beyond have shown, professional conjurers are in general better equipped than scientists to judge the psychic. That said, some scientists—such as Michael Faraday—have shown themselves perfectly capable of being objective about the paranormal.

But others have fallen for it hook, line, and sinker. Robert Hare, the inventor of the oxy-hydrogen blowpipe and one of the preeminent chemists of the nineteenth century, was originally highly skeptical about Spiritualism. In 1853, however, he was persuaded to attend a séance, and his opinion swung to the diametric opposite. In his book *Experimental Investigation of the Spirit Manifestations* (1855), he described how he soon became sure that the rappings, tappings, and table-tiltings could not be the product of trickery.[1] Hare even invented gadgets to be used in communication with the dead that he believed precluded any possibility of cheating. When his own dead father, mother, and sibling promptly started communicating with him, he was a complete convert—and concluded that he himself had mediumistic powers. Among the spirits with whom he thought he communicated thereafter were those of Isaac Newton, George Washington, Lord Byron, Benjamin Franklin . . . All in all, a surprising number of human history's most distinguished figures were attracted from the farthest reaches of the afterlife to this single spirit medium, Hare. It's hardly surprising that his scientific colleagues looked askance at the researches he conducted in his twilight years.

When Hare gave a lecture at the Broadway Tabernacle in November 1855 with the title "Celestial Machinery," he drew an audience of some three thousand, according to the *New York Times*. This was far more than he might have been expected to pull had he stuck to

---

[1] One of the reasons he dismissed fraud as a possibility was because the participants were obviously such an honest lot. Oh, to have had a bridge to sell him.

his earlier passion—chemistry. In fact, he told the audience that his researches into orthodox science were as nothing in comparison to his discoveries about the spirit world.

ALFRED RUSSEL WALLACE

It can be hard to understand why the brilliant naturalist Alfred Russel Wallace embraced Spiritualism with such ardor and persistence. Where other scientists retained at least the pretense of objective detachment, Wallace seemed to swallow the whole kit and caboodle, often being caught in the position of publicly defending the manifestly indefensible—and apparently relishing this role. In the 1876 trial of the fraudulent slate-writing medium Henry Slade (page 40), for example, Wallace appeared for the defense, even though there could be no doubt that the man was a charlatan. How could Wallace be so gullible?

The answer is that, in a sense, he wasn't. He'd lived among "primitive" societies in the Pacific that were organized around belief in the supernatural. Wallace was all for science rising to its rightful status in society, but felt powerfully that it should not do so at the expense of religion. While he didn't feel the actual existence of a god or gods was essential to the well-being of a society, he did think that belief in a god, complete with all the concomitant moral and ethical codes, was vital if Western society were not to devolve into a new form of barbarism. It was the

This statue shows Alfred Russel Wallace, who with Darwin derived the principle of natural selection, in the process of capturing a Wallace's golden birdwing butterfly. His efforts at capturing the truth behind Spiritualist phenomena were less successful.

duty of scientists like himself, Wallace concluded, to ensure that society didn't completely discard the spiritual.

In this context, the best interpretation of his behavior vis-à-vis Spiritualism may be that Wallace deliberately made himself gullible. Certainly it wasn't a matter of his publicly professing belief in what he privately knew to be tommyrot—he was far too sincere a man for any such duplicity. But it seems that he was able to perform the mental trick of persuading himself to accept what was, to the rational mind, preposterous. That same sincerity, moreover, seems to have led him to underestimate the risks of fraud and the lure that the lucrative profession of medium might have.

Wallace was much influenced by his experiences with the medium Agnes "Lizzie" Nichol.[2] Accounts of the manifestations Nichol produced for Wallace and his friends give the impression that she was a clever charlatan; her stunts bear all the hallmarks of extravagant stage conjuring tricks. But Wallace was convinced and reacted with angry defensiveness when critics suggested that she might be fabricating. One of her feats that particularly impressed him was, in the darkness of the séance room, to raise both herself and the chair in which she was sitting from the floor to the top of the table around which her audience sat. It's the kind of stunt a good conjurer might think to devise; why any departed soul might expect it to facilitate communication with the living is more of a mystery.

In 1866, early in his investigations of the psychic, Wallace published a monograph on the subject, *The Scientific Aspect of the Supernatural*. Reading it today is an experience that's both sobering and embarrassing. Clearly the writer is intelligent, but, at the same time, the tract is filled with the same type of contorted

---

[2] Later known, after she married the wealthy Spiritualist Samuel Guppy, as Agnes Guppy and then, after his death and her remarriage, as Agnes Guppy-Volckman.

rationalizations that today you'd expect to find from science-denialist Internet trolls. A sample:

> *But it may be argued, even if [spirits] should exist, they could consist only of the most diffused and subtle forms of matter. How then could they act upon ponderable bodies, how produce effects at all comparable to those which constitute so many reputed miracles? These objectors may be reminded, that all the most powerful and universal forces of nature are now referred to minute vibrations of an almost infinitely attenuated form of matter; and that, by the grandest generalisations of modern science, the most varied natural phenomena have been traced back to these recondite forces. Light, heat, electricity, magnetism, and probably vitality and gravitation, are believed to be but "modes of motion" of a space-filling ether; and there is not a single manifestation of force or development of beauty, but is derived from one or other of these. The whole surface of the globe has been modelled and remodelled, mountains have been cut down to plains, and plains have been grooved and furrowed into mountains and valleys, all by the power of etherial heat vibrations set in motion by the sun . . .*

He can't be blamed for being wrong about the ether—which, before the Michelson-Morley experiment, was widely believed by physicists to permeate the cosmos. But his claim that the bangs and clatters of the séance room could be caused by incorporeal beings because "if electricity and magnetism can do things, so can spooks," is specious thinking, indeed. At some level he must surely have recognized this himself.

In 1867–1868 Wallace tried to persuade the eminent scientists John Tyndall, W. B. Carpenter, G. H. Lewes, and even T. H. Huxley to attend a series of séances with him. Lewes and Huxley seem to have turned him down flat; Tyndall and Carpenter attended one séance apiece, but never again.

SIR WILLIAM CROOKES

Knighted in 1897, Sir William Crookes was a UK physical chemist whose name is commemorated in such items as the Crookes tube and the Crookes radiometer. He's best known in scientific circles for the spectroscopic discovery of the element thallium (1861), for identifying the first known sample of helium (1895), and for being one of the first scientists to investigate plasmas. Following the death of his younger brother, Philip, in 1867, Crookes became an ardent devotee of Spiritualism, attending séance after séance in hopes of contacting his deceased sibling. After the London Dialectical Society set up a committee in 1869 to investigate the possibility of communicating with the dead, Crookes's investigations began in earnest. (He would much later serve as president of the SPR, from 1896 to 1899.)

Sir William Crookes was arguably the most distinguished scientist of his time to be bamboozled by fake mediums.

Among others invited to participate in the London Dialectical Society's probe was the great biologist T. H. Huxley, widely known as "Darwin's Bulldog," the man who demolished the arguments of Bishop "Soapy Sam" Wilberforce in that famous 1860 Oxford debate about evolution. The terms of Huxley's refusal have been widely quoted:

> *If anybody would endow me with the faculty of listening to the chatter of old women and curates in the nearest cathedral town, I should decline the privilege, having better things to do. And if the folk in the spiritual world do not talk more wisely and sensibly than their friends report them to do, I put them in*

the same category. The only good argument I can see in a
demonstration of the truth of "Spiritualism" is to furnish an
additional argument against suicide. Better to live a crossing-
sweeper than die and be made to talk twaddle by a "medium"
hired at a guinea a séance.

Even so, in 1874 Darwin managed to persuade Huxley to attend a
séance and report back on what he saw there. That was his last flirtation with the matter, as he later wrote: "I gave it up for the same
reason I abstain from chess—It's too amusing to be fair work, and too
hard work to be amusing." As for Darwin, he concluded on the basis
of Huxley's report that "an enormous amount of evidence would be
requisite to make me believe in anything beyond mere trickery."

From late 1873 onward, Crookes did much of his psychic investigations using the medium Florence Cook, to the extent that there
were persistent rumors that he and the sixteen-year-old were having
an affair. Florence had been trained in mediumship by two renowned
phonies, Frank Herne and Charles Williams, and had herself been
caught cheating several times before she came to Crookes's attention.
Even so, he was very protective of her, as an incident in 1880 showed.
One of Florence's feats involved her sitting in an enclosed cabinet in
a trance; after a while a white-clad "spirit" called Marie would emerge
from the cabinet into the darkness of the barely lit room and wander
among the sitters, who could even touch her to assure themselves of
her solidity. On this occasion one of the sitters, Sir George Sitwell,
grabbed Marie and held her tight until the lights were turned up—at
which point she was revealed as the medium herself, dressed in white
undergarments. When the cabinet was opened, her outer clothes
were discovered.

Florence maintained that none of this obvious cheating was her
fault: She'd been in a trance throughout the proceedings, so the spirits
must have taken over her body and made it do these things. And

Crookes, her champion, recounted strenuously a parallel incident in 1873 when a sitter, William Volckman (who later married the rival medium Agnes "Lizzie" Guppy, née Nichol—page 13), grabbed the "spirit" that Florence was channeling—in those days known as "Katie King." A struggle ensued and Volckman wound up with a bloody nose, utterly convinced that Katie and Florence were one and the same. According to one observer, the spirit escaped the sitter's clutches by dematerializing her legs and feet, allowing her to wriggle upward and away. When the cabinet was opened, the story went, Florence was found within, in some disarray but fully clothed, with the precautionary seals that had been placed on her still intact. Katie's white robe had vanished as comprehensively as Katie had, and Volckman was denounced for his ungallant behavior.[3]

In another séance, Crookes himself had held Katie and, finding her much like a mortal woman, had asked if he could check the cabinet. She agreed and, when they opened the cabinet together, Florence was allegedly found inside. According to Crookes, this was conclusive proof of Florence's integrity and mediumistic abilities; others pointed out more prosaically that Florence had a sister called Katie who also practiced as a spirit medium.

If the sister was the secret of the feat, this negated a stratagem Crookes had devised to ensure that the "spirit" wasn't just Florence herself. Before each séance he put colored dye on her hands; when Katie had emerged from the cabinet, it was easy enough, even in the near darkness of the chamber, to check her hands. It's some gauge of his level of scientific objectivity that, when he used the same technique with a similar but far less known medium—the amateur Mary

---

[3] Volckman's version was a tad different. He said that, as he held the struggling spirit, the lights were suddenly extinguished and a burly stranger wrenched the ethereal form from his arms . . . taking some of Volckman's beard hairs with it! Volckman was then expelled from the séance room.

⇒ EMPYREAN CIGARS AND CELESTIAL SCOTCH ⇐

Knighted in 1902, Oliver Lodge was a physicist and one of a number of distinguished siblings, among them the mathematician Alfred Lodge and the historians Eleanor Lodge and Sir Richard Lodge. His major scientific accomplishment was his demonstration in 1893 that the luminiferous ether did not in fact exist; this work was an important precursor for the development by Einstein of relativity theory. He also pioneered wireless telegraphy.

Despite such distinction, Oliver Lodge is probably best remembered today for his investigations of Spiritualism and paranormal phenomena. These began in the late 1880s but became ever more intense after 1915, when his son, Raymond, was killed in battle. Before Lodge died, he promised that he would make the truth of his claims about life after death manifest by returning to show himself. Alas, no such sightings were ever recorded.

In *Raymond, or Life and Death* (1916), Lodge's son, channeled through his father, described the afterlife in dishearteningly mundane terms: The dear departed could smoke cigars and drink scotch if they so chose, while most—despite a nostalgic yearning for the garb of their terrestrial lives—went around in white robes. Some devoted their energies to working in scientific laboratories to manufacture useful items . . .

Oddly, although Lodge could accept the possibility of life after death, he refused to accept the possibility of Einstein's relativity. At a joint gathering of the Royal Society and the Royal Astronomical Society in 1919, where it was announced that the theory had been confirmed and all future physics must accept it, Lodge stormed out.

Rosina Showers—and discovered the "spirit" had colored hands, he immediately concocted the "explanation" that this was only to be expected because the "spirit" had drawn its substance from Showers's body and so could inadvertently have brought along the dye as well. Had it not been for the fact that the garrulous Showers confessed her trickery to another medium, who promptly informed Crookes, he might have made a very public fool of himself.

As it was, he had a secret meeting with Showers, during which she promised never to cheat again while he promised not to expose her for having done so. Showers's mother, learning of the secret meeting, assumed the worst and spread the rumor that Crookes was in the habit of investigating more than just the psychic abilities of female mediums. Of course, Crookes couldn't convincingly refute this tittle-tattle without breaking his promise to Showers. It seems that this was what finally made him decide to give up psychic investigation for good.

Before that, however, Crookes had had the chance to evaluate Anna "Annie" Eva Fay, an American medium who visited London in 1875 and played to large and enthusiastic audiences. The stage magicians John Nevil Maskelyne and George Alfred Cooke put on a rival demonstration in which they duplicated her "psychic" feats using conjuring, which somewhat stymied her career prospects in the United Kingdom. Indeed, throughout her career she was regularly caught cheating; on the other hand, she was a very pretty woman, which might have led some people to give her the benefit of the doubt. Crookes, certainly, having tested her using an electrical gadget he'd developed to test Florence Cook, judged her to be genuine. And, since he was a Fellow of the Royal Society, she billed herself thereafter as endorsed by the Royal Society!

Despite all this, Crookes wasn't entirely gullible. He did differentiate in his own mind between those séances he attended on the premises of the medium, whether it be Daniel Dunglas Home,

Florence Cook, or Kate Fox, and those—far fewer in number—where the medium could be persuaded to come to Crookes's own home. Clearly, the former were far more susceptible to chicanery and advance preparation than the latter, so it was only the results from the latter that Crookes included in his "scientific" papers on the subject. At the same time, though, he was hardly the dispassionate experimenter he presented himself to be.

❧ THE SOCIETY FOR PSYCHICAL RESEARCH AND ITS BRETHREN ORGANIZATIONS

The Society for Psychical Research (SPR) was founded in 1882 by a group of dons from Trinity College, Cambridge, in an attempt to mount a scientific investigation into the psychic phenomena that were so popular in Europe and North America at the time. Among the founders were physicist Sir William Barrett (arguably the prime mover), psychologist Edmund Gurney, philologist F.W.H. Myers, journalist Edmund Dawson Rogers, and economist Henry Sidgwick. The first volume of *Proceedings of the Society for Psychical Research* was issued in 1883; the *Journal of the Society for Psychical Research* first appeared in 1884.

The American Society for Psychical Research (ASPR) was founded soon after its UK counterpart, in 1885. Among its founders was the psychologist and philosopher William James, and its first president was the astronomer Simon Newcomb.[4] This first ASPR unraveled shortly after the death of its then-director, Richard Hodgson, in

---

[4] James would serve as the (UK) SPR's president in 1894–1895. Newcomb was skeptical about Spiritualism, as he was about most things. Among other judgments from him are: "Flight by machines heavier than air is unpractical and insignificant, if not utterly impossible" and "We are probably nearing the limit of all we can know about astronomy"!

1905. But in 1906 a new ASPR arose in New York from the ashes of the old, under the initial control of the Columbia ethics professor and psychic investigator James H. Hyslop. The first volume of the *Journal of the American Society for Psychical Research* was published in 1907.

One of the ASPR's benefactors was Chester Carlson, inventor of the Xerox process, who contributed funds while alive and, on his death in 1968, left the organization $1 million in his will. (He also left another $1 million to the University of Virginia School of Medicine for the creation of a Division of Parapsychology.) A further major bequest came the ASPR's way in 1972 from the estate of James Kidd, an Arizona prospector who'd gone missing as long ago as 1949. It was assumed that he'd been a poor man, but in 1956, after he'd been declared legally dead, it was discovered that he'd managed to accumulate in excess of $270,000 (about $2.7 million today) in stocks. His will instructed that his fortune "go into research or some scientific proof of a soul of the human body which leaves at death I think in time their [*sic*] can be a photograph of soul leaving the human at death." Some 130 Spiritualist organizations petitioned the court, each claiming to be the ideal body to carry out Kidd's wishes, and the court battle—dubbed the "Soul Trial" by the press—went on for years. Among the frontrunners were the ASPR, the Barrow Neurological Institute at St. Joseph's Hospital in Phoenix, and the Psychical Research Foundation of Durham, North Carolina. In the end, Judge Robert L. Myers opted for the Barrow Neurological Institute but, on appeal, the Arizona Supreme Court awarded the legacy instead to the ASPR, with about one-third of it going to the Psychical Research Foundation. Ironically, none of the money was spent on trying to photograph the soul.

The French branch of the SPR, La Société Française pour Recherche Psychique (SFRP), came along in the same year as the ASPR: 1885. The Scottish branch of the SPR was founded as recently as 1987. There are other equivalent organizations around the world.

Among the SPR's early presidents were some very distinguished men, including:

- Arthur Balfour, later UK prime minister
- Scottish folklorist Andrew Lang
- English philosopher C. D. Broad
- American psychologist William James
- English physical chemist Sir William Crookes
- English physicist Sir Oliver Lodge
- French physiologist Charles Richet
- English physicist Lord Rayleigh

Even up to more recent times, some very respected scientists have served as SPR presidents. In other words, the SPR is no dodgy

An 1885 print showing some of the Fellows of the Royal Society. Spiritualism divided the scientific world. Both Lord Rayleigh (standing, center) and Sir William Crookes (seated, far right) served as presidents of the SPR. T. H. Huxley (seated, fourth from left), by contrast, made his unbelief in the psychic bluntly known. E. Ray Lankester (standing, fifth from left) was a prominent debunker of psychic fraud (see page 39).

fly-by-night organization populated solely by the credulous—although it has to be accepted that, over the decades, its membership has included plenty of the credulous among the skeptics. Prominent nonscientists actively involved with the SPR in its early days included author Mark Twain, art critic John Ruskin, poet laureate Alfred, Lord Tennyson, philosopher Henri Bergson (who served as the Society's president in 1913), former UK prime minister W. E. Gladstone, and painter Sir Frederick Leighton.

At the time of the SPR's founding, many Spiritualists looked on it askance or with intense dislike. The SPR's intention of bringing the scientific method to bear upon the psychic field was clearly a threat to many of its practitioners. It's important to note that not all the promoters of the SPR and the various related societies were convinced that the paranormal actually existed; what they supported was research to find out whether or not it did.

In her excellent history, *The Spiritualists* (1983), Ruth Brandon asserts that, despite its scientific bent, the real impulse behind the formation of the SPR was religious. After the publication in 1859 of Darwin's *On the Origin of Species*, it was difficult for educated people to believe in such Judeo-Christian tenets as the Creation. While the devout could accept evolution intellectually as God's way of doing things—they had, after all, come to accept the geologists' uniformitarian theory of Earth as an exceptionally old planet—it nevertheless seemed as if religious faith as a whole might be about to perish as a tenable worldview, and with it such spiritual concepts as man's immortal soul.

People like Sidgwick, Gurney, and Myers were hoping science would confirm that there was still something spiritual there to which they might cling. Writing in 1880, Sidgwick put the dilemma succinctly: "*Either* one must believe in ghosts, modern miracles, etc., *or* there can be no ground for giving credence to the Gospel story." Much later the distinguished SPR researcher Eric Dingwall suggested that Myers and other members had a different agenda from both this and the SPR's stated aim of scientific research: "Myers, among others . . . knew that the primary aim of the Society was not objective experimentation but the establishment of telepathy."[5] If this is true, it casts a new light on the history of the SPR![6]

Although initially focused largely on Spiritualism, apparitions, and phantasms, the SPR's interests shifted during the twentieth century to such matters as survival after death and ESP. One idea that became popular among the researchers was that the entities previously regarded as mediums' spirit guides or controls might instead be manifestations of multiple personality disorder (now called "dissociative identity disorder"). The focus shifted from the notion that psychic phenomena were being produced through communication with those in the afterlife toward explanations that operated in terms of ESP and, especially, psychokinesis. Somehow faith in psychic phenomena was regarded as less outlandish and more respectable than faith in the afterlife.

THE CROSS-CORRESPONDENCES

During the heyday of the SPR, the most significant single piece of experimentation concerned what came to be called "the cross-correspondences." Various SPR members had promised they'd try to communicate posthumously with surviving colleagues, and after their deaths several mediums had indeed claimed to have been contacted. But the SPR's researchers were well aware of how easily such contacts could be faked. What was needed was some more complex form of message, transmitted through more than one medium, that would only make sense when the disparate communications were all put together.

---

[5] "The Need for Responsibility in Parapsychology," in Allan Angoff and Betty Shapin, eds., *A Century of Psychical Research*, 1971. The essay can be found online at www.SurvivalAfterDeath.info.

[6] Another phenomenon considered by the SPR to be within its purview in its early days was hypnosis.

At the focus of the cross-correspondences were four automatists:[7] "Mrs. Willett" (in reality the suffragist Mrs. Winifred Coombe-Tennant), "Mrs. Holland" (in reality Mrs. Alice Fleming, sister of Rudyard Kipling), Helen Verrall, and Helen's mother, Margaret Verrall. All told, the case involved some three thousand automatically written scripts produced over about three decades. Only when these were analyzed together in detail was it possible to see the pattern as a whole—the picture being built up by countless tangential allusions. But that could also be where the case is misleading. It's easy for any of us to fall into the trap of a sort of

A tragic love affair of Arthur Balfour (shown here in 1904), who served as UK Prime Minister between 1902 and 1905, served as the focus of the cross-correspondences experiment.

intellectual pareidolia, where every piece of evidence we come across seems to fit in precisely with our preconceived conclusion. Show Erich von Däniken a thousand Peruvian carvings, and he'll demonstrate how each of them must depict an ancient alien visitation. Similarly, show three thousand pieces of automatic writing to someone convinced of the reality of psychic communication, and he or she is likely to see a pattern emerging that supports his or her preconception.

The incident at the heart of the cross-correspondences was the death from typhus in 1875 of Mary Catherine Lyttleton, with whom

---

[7] Automatism, or automatic writing, refers to the system whereby the medium sits, pen in hand and poised over paper, and allows "the spirits" to guide the hand as it writes messages. Later it came to be regarded as having less to do with the spirits than with psychokinesis, and today it's generally seen as a product of the ideomotor effect—see page 53.

the young Arthur Balfour was deeply in love. Balfour, an intensely private man for a public figure, had never told anyone at the time or later—even his family—of this love, or that he and Mary had been on the brink of betrothal when she fell ill and died. Many of the scripts that make up the cross-correspondences feature the endeavors of Mary, assisted by various sympathetic spirits, to get in touch with Balfour and tell him that she still loved him. Since the automatists, at least in the early days of the experiment, knew nothing of the liaison, they were uncertain as to who was trying to get in touch with whom; only later did the story emerge.

This, of course, is to assume all four automatists hadn't before-hand established contact with each other. The Verralls obviously could hardly do otherwise, but at the outset of the affair Alice Fleming was with her husband in India. As Janet Oppenheim concludes in her history of English psychic research from 1850 to 1914, *The Other World* (1985):

> *It is possible that collusion, fraud, and self-delusion played their part in the [cross-correspondences] case, as in so many other instances in psychical research, but, for once, it is highly improbable, given the number of people, the sheer volume of material, and the span of time involved.*

The factor that Oppenheim seems to ignore here is the role of the interpreters of the assembled messages. As noted, the precon-ceptions of those investigators had the potential to affect very significantly the meaning that was derived from this "sheer volume of material."

Balfour himself did not learn of the cross-correspondences until 1916, about halfway through the experiment. By the time of his death in 1930, it seems he was persuaded that he was on his way to rejoin his beloved Mary.

❧ FAUX MEDIUMSHIP

"Take the séance room out of spiritualism and you reduce it to another drab religion."

—M. Lamar Keene, *The Psychic Mafia* (1976)

Mediums have been around since long before the Spiritualist movement. In earlier days, as people having the power to raise and communicate with the dead, they were known as necromancers in some cultures, shamans in others. We could certainly make a case that the spirit medium and/or psychic fulfills the same sort of role in modern societies as the shaman did in older ones, and many self-professed psychics may be attracted to the profession simply to have the opportunity to exert power over other people, much as shamans did. It's hard, otherwise, to find a plausible motivation for supposed psychics like Sylvia Browne to tell parents—on television—that their children were dead, when in fact she had no way of knowing whether or not this was true. It's an act of unspeakable cruelty, but, in its pathetic way, it's also an expression of power—not quite of life and death, but something close to that.

The (relatively) early days of the SPR were bedeviled by bogus mediums. It had been in part because of the Creery sisters—the five daughters of the Reverend A. M. Creery whose seeming telepathic feats had impressed Sir William Barrett, Edmund Gurney, and F.W.H. Myers—that the SPR had been founded. When the girls were caught cheating, they maintained they'd only done it once or twice.

The experiments the researchers did with the Creery sisters were based on an old parlor game. One girl left the room while the others picked a card or wrote down a word or number. When the girl returned, the others would "wish" the relevant information at her as hard as they could until, sooner or later, she "received" it and could call out the word or the number or the identity of the card. As we might guess, it eventually emerged that the girls were using a simple

Eusapia Palladino levitates a table in Milan in 1892, with psychic investigator Alexandre Aksakof invigilating.

code—not too dissimilar from the ones used by stage "mentalist" acts. The researchers might have caught on sooner had they not been so impressed by the integrity of the good Reverend Creery, whose delightful children surely could not even contemplate duplicity.

Although the SPR followed standard scientific practice—if you catch someone cheating, you automatically regard all the rest of that person's work as suspect, at best—and so discounted the Creery case as bogus, Barrett objected strongly, claiming that, while the Creery sisters had later resorted to cheating, the early experiments with them were genuine evidence of thought transference. Oddly, in the 1930s the American researcher J. B. Rhine took Barrett's side in this, accepting the experiments with the Creery sisters as strong evidence for ESP.

Similarly, as late as 1997 parapsychologist Stephen E. Braude, in his book *The Limits of Influence*, was taking seriously the cases of Daniel Dunglas Home (page 32) and Eusapia Palladino (page 29). Braude seems to have been discontented with the scientific method as a whole. His opening chapter is titled "The Importance of Non-Experimental Evidence," and in his preface he laments:

> *Not too long ago, before I began to investigate the evidence of parapsychology, I still believed that . . . my colleagues in academia (especially in philosophy and science) were committed to discovering the truth, and that intellectuals would be pleased to learn they had been mistaken, provided the revelation brought them closer to this goal. I now realize how thoroughly naive I was.*
>
> *Since dipping into the data of parapsychology I have encountered more examples of intellectual cowardice and dishonesty than I had previously thought possible. . . .*
>
> *I have seen how scientists are not objective, how philosophers are not wise, how psychologists are not perceptive, how historians lack perspective—not to mention (while I'm at it) how physicians are not healers . . .*

The supposed telepathist Douglas Blackburn, who with his accomplice, George Albert Smith, had been the subject of a protracted investigation by SPR cofounder Edmund Gurney, admitted in public some two decades after Gurney's death[8] that he and Smith had faked all the experiments. Smith, who was still working for the SPR, vehemently denied the charges but, as with the Creery sisters, the Society decided the Blackburn-Smith evidence could no longer be relied upon.

The most damaging confession of all was that of the Fox sisters (page 6), but there were other revelations that could have been equally fatal. The medium Eusapia Palladino was deemed genuine by scientists of the caliber of Sir Oliver Lodge and Charles Richet and by

---

[8] Gurney's death may have been a suicide. An alternative explanation is that he was in the habit of sniffing chloroform to numb the pain of his chronic neuralgia and accidentally overdosed.

criminologist Cesare Lombroso, yet she was caught cheating not just from time to time but *often*. In fact, she was downright clumsy, almost as if she wanted to be caught. One excuse offered was that, if she was such an incompetent cheater, she must genuinely have been channeling spirits all the times when she *wasn't* caught cheating!

Palladino, by the way, was quite open about her sexuality and the fact that she found the trance state erotic—often to the point of orgasm. In the Victorian era, the sound of the medium in the throes of sexual passion must have served as an excellent distraction to would-be objective investigators. Some researchers, Lombroso among them, speculated that her mediumistic powers might be some form of sexual sublimation.

Not long after her death, Palladino returned—this time speaking through a promising young Neapolitan medium named Nino Pecoraro. Pecoraro greatly impressed Sir Arthur Conan Doyle and his wife, Jean—but, then, it was hard to find a medium who didn't—and, more seriously, the investigating committee for a 1923 *Scientific American* contest offering a $2,500 prize to anyone who could convince the "experts" that they were genuinely psychic. Tightly tied up, Pecoraro produced all sorts of psychic manifestations—strange sounds, flying objects, the works. He passed the first two tests with flying colors, but illusionist Harry Houdini joined the committee for the third test and personally supervised the preliminary tying-up of the medium. According to the *New York Times*'s December 19, 1923, issue:

> *It took almost an hour and three-quarters to prepare the medium for his test, which was probably the most severe he ever underwent. His hands were sewn into gloves attached fast to his underclothing. His hands were then placed, Chinese fashion, into the sleeves of his coat and sewn there. The sleeves were sewn to the coat itself and the coat to the trousers, both in front and behind.*

> *Then Houdini put the boy in a chair and tied him with*
> *short pieces of rope which won even the admiration of the*
> *"spirit," which referred to his success time and again during*
> *the sitting. The chair was put into the cabinet, which was*
> *formed by black curtains hung in one corner of the room, and*
> *screwed to the baseboard of the room with metal bands,*
> *which were sealed.*

Unsurprisingly, the spirits failed to do much by way of manifesting that night. The *Times* continued:

> *It was only a few minutes after the lights were put out that the*
> *manifestations began, with creaks from inside the cabinet.*
> *There were sounds of an extra chair in the cabinet being tilted*
> *back and forth throughout the séance, and at times, a*
> *tambourine and bells on the chair could be heard. . . . At the*
> *outset "Palladino" complained of the way they had tied Nino,*
> *but she promised phenomena in spite of it. Little, however,*
> *became manifest.*

The committee seems to have bent over backward to be charitable in their conclusions. They deduced that there was fraud going on, but declared themselves certain that Pecoraro was going into a trance. The perpetrator of the fraud, then, was not Pecoraro, but rather Pecoraro's subconscious!

And the list goes on. Another cheating "psychic" associated with the SPR was Helena Blavatsky, cofounder of the Theosophical Society. In 1885 the veteran SPR researcher Richard Hodgson reported in the Society's *Proceedings* that, following extensive interviews with and demonstrations by Blavatsky's housekeeper Emma Coulomb and Coulomb's husband, it was patently obvious that Blavatsky—publicly renowned for her ability to produce physical

Helena Blavatsky, queen of the phonies, 1889.

"apports" (objects out of thin air)—had been cheating on a grand scale. As a result of this report, Blavatsky—who, with her devoted disciple, Colonel Henry Olcott, had set herself up in some style in India as a guru—fled back to London where, implausibly, her reputation survived.

At some point the SPR introduced a policy stating that, if ever a medium was discovered cheating—even just once—the Society's researchers would never work with that medium again. In her 1982 book *The Society for Psychical Research: A History 1882–1982*, Renée Haynes laments that this policy "probably deprived [the SPR] of much valuable material, since . . . some fakes were undoubtedly used to fill in when the faculty for producing the genuine article had dried up." What Haynes neglects to recognize is that the policy—besides being in accord with the principles of scientific research—saved the SPR's files from becoming clogged up with a vast amount of "noise" in the form of fakery that would have made it nearly impossible to separate out any material that might truly be of worth to their cause. If you have a hundred false reports and three true ones, the three true ones might as well not exist.

❧ THE HOVERING HOME

One of the most celebrated mediums of all time, Daniel Dunglas Home, was born in Edinburgh in 1833, adopted in infancy by an aunt, and reared in Connecticut from the age of nine. In 1855 he returned to the United Kingdom for a first brief visit, having supposedly left his Connecticut home because his dramatic parapsychological manifestations had persuaded the rest of the family that he might be possessed by Satan. (Other sources more prosaically suggest that he was sent abroad to help him recuperate after a bout of tuberculosis.)

The United Kingdom was ripe for Spiritualist mediums, the craze having spread from the United States and taken hold rapidly. Home spent the next few years on the European mainland, where he performed séances for wealthy patrons, and then returned to England in 1859, following the birth of his son. (He had married a young Russian noblewoman in St. Petersburg, his best man being the writer Alexandre Dumas.)

Unlike almost every other medium of his era, Home was never publicly exposed as a cheat, although several of his sitters reported privately that they'd noticed apparent shenanigans. There's ample circumstantial evidence that he cheated, however, in that (a) his powers tended to ebb and flow according to the strictness of the scientific controls in place during a particular test and (b) the spirits stubbornly refused to cooperate whenever there might be a professional magician in attendance. And, again unlike other mediums of his times, Home declined to charge fees for his sittings—although he did accept gifts. Those gifts poured in with sufficient generosity that he became a rich man. His clients in the UK and Europe included members of the aristocracy and even royalty; among them was the writer Sir Edward Bulwer-Lytton, whose novel *A Strange Story* (1862), which devotes an entire chapter to a debate between a medium and a skeptic, may owe more than a little to the author's experiences with Home.

Although Home did all the stuff the other mediums did—the rappings, the spirit messages, and the like—his specialty was levitation: of furniture, of people, and, most especially, of himself. It's frequently stated in the Spiritualist/paranormal literature that he performed these prodigious feats of levitation in a brightly lit environment, but closer examination of contemporaneous accounts reveals that this was not so. While the room was well lit when his séances began, by the time the self-levitation started the lights had been turned down low enough that all the sitters could make out were vague shapes and, passing in front of dimly lit windows, silhouettes. Home would obligingly mark the ceiling with a cross to "prove" he had floated up there—a stratagem he wouldn't have needed to employ had he simply left the lights on. All his feats were within the scope of a competent illusionist.

The best known of Home's illusions occurred one evening in December 1868 at Ashley House in London when, to the astonishment of a trio of distinguished witnesses—Lord Adare (the earl of Dunraven), Lord Lindsay (the earl of Crawford), and a Captain Wynne—Home apparently floated out through a third-floor window and then back in through another. The event seems to have been a masterpiece of misdirection by the medium. He told his three visitors what they were about to witness and then gave them enough visual clues (in a darkened room, of course) for them to be able to piece together a false narrative afterward. Each man saw only bits of the "levitation"; it seems most likely that Home simply stepped out one window, crept along a ledge or a plank between the two window balconies, and then stepped back into the room.

The venom in Robert Browning's famous epic verse about Home, "Mr. Sludge, 'The Medium'" (in *Dramatis Personae*, 1864), derived in large part from the fact that the medium's eye had been caught by the poet's wife, Elizabeth Barrett Browning. Although she didn't respond to the much younger Home's advances—despite being mightily impressed by his mediumship—the situation created tension in the Browning household. When, after Elizabeth's death, Home called by with a sort of "Let bygones be bygones" proposal,

Browning threatened him with physical damage. The poem, which skewers Home for what Browning could see was outright chicanery, opens thus:

> *Now, don't sir! Don't expose me!*
> *Just this once! This was the first and only time, I'll swear,—*
> *Look at me,—see, I kneel,—the only time,*
> *I swear, I ever cheated,—yes, by the soul*
> *Of Her who hears—(your sainted mother, sir!)*
> *All, except this last accident, was truth—*
> *This little kind of slip!—and even this,*
> *It was your own wine, sir, the good champagne . . .*
> *Which put the folly in my head!*

Throughout, there's little question that the "sludge" of the title was intended by Browning as a euphemism.

In 1869, Home was subjected to a series of four tests by the London Dialectical Society. To the disappointment of the investigating committee, led by Dr. James Edmunds, he failed to produce any levitations or, indeed, any other manifestations that Edmunds himself wasn't able to duplicate using strictly nonpsychic means. In 1870, Home was tested at the University of St. Petersburg, and again the result was a damp squib. One manifestation the Russian scientists were keen to investigate was the "spirit hand" that

The poet Robert Browning, shown in this 1888 photo, was not at all amused by Daniel Dunglas Home's interest in his wife, Elizabeth.

sometimes visited Home's private séances. Skeptics suggested the "hand" was in actuality one of the medium's feet, slipped out of his shoe and brought up onto the tabletop. (Home was a very athletic, limber man.) So the Russians set him to work at a glass-topped table and, sure enough, the "spirit hand" declined to appear.

Sir William Crookes became interested in Home through reading *Experiences in Spiritualism with Mr. D. D. Home* (1869) by the earl of Dunraven. (Ironically, after writing his book Dunraven lost interest in Home and in Spiritualism in general.) Crookes contacted Home in 1869 and within the year published the first of four articles on Spiritualism in the *Quarterly Journal of Science*—which, conveniently, he himself owned and edited. The last of these four articles, "Notes of an Enquiry into the Phenomena called Spiritual during the Years 1870–1873," summarizing his Spiritualist investigations, appeared in the journal's January 1874 issue. (After that year, having failed to persuade his scientific colleagues of the reality of Spiritualism, Crookes focused once more on his orthodox scientific studies. Toward the end of his life, however, his interest in the psychic was reawakened as he attempted to communicate with his dead wife, Ellen.)

In an early experiment with Home, Crookes set up a length of mahogany with one end on a table and the other supported by a spring balance. Crookes and a colleague, the eminent astronomer William Huggins—who much later served as president of the Royal Society—observed as the spring balance registered several pulses of downward force that ranged up to six pounds (2.7 kg), although Home's fingers barely touched the tip of the mahogany plank at the table end. By contrast, even by standing on the end of the board, where Home's

An 1874 etching of Daniel Dunglas Home's accordion, which he claimed to be able to play from a distance.

fingertips had merely rested, Crookes couldn't get the indicator to register more than about two pounds (1 kg). Crookes and some of the others present felt it was impossible for those pulses to represent anything other than a psychic force. Huggins was significantly less convinced, later writing to Crookes, "The experiments appear to me to show the importance of further investigation, but I wish it to be understood that I express no opinion as to the cause of the phenomena which took place."

Huggins was, of course, correct in his caution. The SPR's Frank Podmore, writing much later on the subject in his *Modern Spiritualism* (1902), observed that "Home—a practiced conjurer, as we are entitled to assume—was in a position to dictate the conditions of the experiment." Moreover, Crookes underestimated the importance of all the occasions on which Home couldn't get the balance to register at all, regarding these as merely instances when the ever-contrary psychic force had deserted the medium. A less charitable interpretation would be that these failures represented times when Crookes and his fellow experimenters were exercising vigilance, rather than letting themselves be distracted.

In 1871, Home gave up the practice of mediumship and went to live in France with his second wife, the heiress Julie de Gloumeline. His sole contribution to Spiritualism thereafter was his book *Lights and Shadows of Spiritualism* (1877), in which he exposed at great length the methods of fraudulent mediums—although not his own!

One of Home's specialties had been the sounding of an accordion that would be placed in a case in the séance room far from anybody's hands, including his own. It would play simple tunes like "The Last Rose of Summer." After his death, found among his possessions were a collection of tiny harmonicas—small enough that they could be kept in the mouth, then manipulated to the lips. There, they'd be almost undetectable behind Home's big Victorian mustache!

❧ SLATE SCAMS

In the 1880s, Richard Hodgson and S. J. Davey of the SPR became interested in using a knowledge of conjuring to expose the fakers. Hodgson had done a sterling job for the SPR in demolishing the claims of Helena Blavatsky, and Davey was a young man in frail health who had taken up conjuring as a way of amusing himself during his long periods of enforced solitude. They were

One of the tricks the slate writers used to persuade gulls that the spirits had left a message was inscribing the message on the slate with their bare feet.

brought together through their separate investigations of the fraudulent slate-writer William Eglinton.[9] Davey was initially fooled by the charlatan but, on reflection, realized he could replicate many of Eglinton's feats through conjuring.

It's not clear if Davey approached the far more experienced Hodgson or vice versa, but the two hatched the scheme of staging fake slate-writing séances as a way of demonstrating to the gullible how easily they could be duped. The séances were reportedly sensational and were certainly very popular. In one of Davey's tricks, he faked slate-writing by rigging up a thimble with a piece of chalk at

---

[9] Slates were sealed such that no mortal hand could write on them, then placed below the séance table. During the séance, sitters would hear a dramatic sound of scribbling. When the sealed slate was opened to view, all could gasp at the messages the spirits had left there. The easiest way to effect this wonder was to swap the blank slate for one the medium had prepared earlier.

its tip. As soon as he started to put the slate beneath the edge of the table, he'd quickly scribble something on the slate's underside. Then he would pull the slate briefly back out on the pretext of demonstrating that its upper surface was indeed blank before deftly turning it over as he slipped it back under the table. Finally—presto!—he'd bring the slate out again to show the writing on its upper face.

In other instances he would rely on a friend of his, Henry Acland Munro, to creep into the darkened séance room and fake spirit apparitions. On many occasions he and Munro were nearly unmasked and, to divert the attention of the sitters, had to improvise hastily and sometimes clumsily. What Davey found was that, remarkably, people tended to banish these glitches and other bits of stage business from their memories of the events. In the slate-writing example above, almost no one recalled Davey pulling the slate briefly back out from under the table.

When finally, in 1887, Davey and Hodgson published an account of what they'd been doing in the SPR's *Proceedings*, some of the more die-hard SPR members took severe umbrage: The job of the SPR, they averred, was to support the existence of the spirit world, not question it! Other assaults on the exposure took a different tack: The mock séances were so impressive that they must be genuine! Although Davey and Hodgson claimed everything was being done by trickery, was it not more likely that Davey was a genuine and very powerful medium, pulling the wool over the skeptical Hodgson's eyes? Alfred Russel Wallace was one to swallow this particular farcical argument, writing in 1885, "Unless all can be explained, many of us will be confirmed in our belief that Mr. Davey was really a medium as well as a conjurer, and that in imputing all his performances to 'trick' he was deceiving the Society and the public."

A more direct approach had been taken a few years earlier by the UK zoologist E. Ray Lankester, infuriated by what he regarded as the outright fraud of this new craze, slate-writing. He chose as his target the most prominent and richest of its exponents, the American medium Henry Slade, then plying his lucrative trade in the UK. At one of Slade's séances in London, Lankester grabbed the slate as soon as it had been placed beneath the table, before the "scribbling" supposedly started, and showed that the writing was already there. Prosecuted for fraud in October 1876 at Lankester's instigation, Slade received a sentence of three months' hard labor, although this was overturned on a technicality. Slade fled to Germany and later spread the myth—readily swallowed by the journalists of his homeland—that he'd escaped imprisonment by demonstrating slate-writing in open court while bound, gagged, and blindfolded.

In Germany, Slade succeeded in bamboozling Karl Friedrich Zöllner during a series of sittings in late 1877. Zöllner, an astrophysicist at the University of Leipzig, invited various prominent scientists to these sessions—including psychologist Wilhelm Wundt, physicists Gustav Fechner and William Weber, and mathematician Wilhelm Scheibner. Samuel Bellachini, court conjurer to the emperor of Germany, declared that he could discover no trace of fraud in Slade's act—somewhat surprising, considering how easily Lankester had found it to trap the charlatan.

From 1878 onward, Zöllner published accounts of the sessions he organized with Slade. On December 17, 1877, for example, he used two prisms and a saccharometer to test Slade's ability to change the polarity of a chemical solution. Slade chose not to change the polarity but rather to use the apparatus as a means of reading a book clairvoyantly. Astonishingly, instead of hewing to his original experiment, Zöllner decided this was evidence that Slade was being aided by beings from the fourth dimension and promptly improvised two experiments designed to test this hypothesis. One of these involved the marking of a sealed slate, and the other entailed the knotting of a cord by psychic means. In April 1878 William Crookes published in his *Quarterly Journal of Science* Zöllner's paper, "On Space of Four

Dimensions," in which Zöllner used the cord-knotting experiment as a basis for his claim that space had a fourth dimension—offering to show people the knotted cord as proof!

In 1878–1879, Zöllner published his massive three-volume *Wissenschaftliche Abhandlungen*, primarily devoted to Slade and today better known under the title of the (slightly abridged) 1880 English translation, *Transcendental Physics*. The furor that greeted the German publication of this book lasted for years. The source of the fuss was the involvement of such prominent scientists in promoting a purveyor of irrationality—moreover, a purveyor who had been shown in the UK courts to be a fraud. Zöllner's death in 1882 did nothing to dampen down the controversy. Significantly, one of the most skeptical commentators was Wundt, who had actually seen Slade in action. While he was not willing to dismiss the paranormal entirely, Wundt was fairly certain that all he had witnessed Slade perform was trickery—for example, the slate-writings were full of the kinds of grammatical errors you'd expect when an English speaker did his best in German.

In early 1885, the Seybert Commission (page 8) investigated Slade and concluded, like the British before them, that the man was a phony. The commission's secretary, George Fullerton, made his own comments on the Zöllner affair after having come to Germany to conduct a personal investigation. Slade, in his own defense, had been able to point to the support his authenticity had received from several very estimable German scientists. In his report Fullerton noted that, at the time of the

The physicist Gustav Fechner, one of the somewhat decrepit team recruited by Karl Friedrich Zöllner to examine the fake slate writer Henry Slade.

séances, Fechner was partially blind, Scheibner's vision was little better, Weber was of extremely advanced age, there was reason to believe that Zöllner himself had been of unsound mind, and Wundt had considered Slade a fraud.

✿ POLEAXED BY PIPER

The medium who really bamboozled the investigators of the SPR was Mrs. Leonora Piper. She persuaded the usual crop of distinguished scientists that she was truly in touch with the dead. Far more impressive is that she persuaded the SPR's veteran researcher, Richard Hodgson, that she was genuine—a difficult feat, since Hodgson had exposed countless other fraudulent mediums who'd fooled his colleagues.

Piper's mediumship supposedly started when, at age eight, the young Leonora Simonds was playing in the garden and suddenly felt as if she'd been walloped on her right ear. There was also a hissing sound that eventually resolved itself into the message: "Aunt Sara, not dead, but with you still." Sure enough, the family later heard that Aunt Sara had died at that very moment.

Johann Sebastian Bach: just one of Leonora Piper's myriad spirit controls.

Things really took off in 1884. By now Piper was married and had had her first child, Alta. She was suffering continuing pains after the birth and, concerned that she might have an ovarian tumor, consulted the blind clairvoyant J. R. Cocke in hopes of obtaining a diagnosis and cure. At their first meeting she felt uneasy; at their second, she fell into a trance and was taken over by the spirit of a young Native American woman called "Chlorine."

Chlorine was only the first of the numerous controls Piper claimed to be working with over the decades. Others in this startlingly eclectic bunch included Johann Sebastian Bach, Julius Caesar, George Eliot, Henry Wadsworth Longfellow, William Stainton Moses (a medium who had apparently arrived on Piper's scene with a bunch of spirit controls he had worked with during his own life), Loretta Ponchini (an Italian girl), Sir Walter Scott, actress Sarah Siddons, Commodore Cornelius Vanderbilt, and especially a French physician named Phinuit, who "adopted" her in 1884 and remained her control until 1892. Interestingly, Phinuit's medical knowledge was at best rudimentary and often just plain wrong— and his French was lousy.

In the spring of 1886, William James published his paper, "A Report of the Committee on Mediumistic Phenomena," in the ASPR's *Proceedings* and appraised Piper thus: "I am persuaded by the medium's honesty, and of the genuineness of her trance . . . I now believe her to be in possession of a power as yet unexplained." Elsewhere he remarked of her, "To upset the conclusion that all crows are black, there is no need to seek demonstration that no crows are black; it is sufficient to produce one white crow; a single one is sufficient." As noted, Richard Hodgson, probably the SPR's most capable debunker of fraudulent mediums, finally met his match in Piper and wholeheartedly endorsed her psychic abilities. This was to the great dismay of his friend Morton Prince, a noted psychologist, who remarked that the matter had "wrecked Dick Hodgson, who had one of the most beautiful minds I ever knew."

In 1892 Piper attracted a new spirit control, Dr. George Pellew, a recently deceased lawyer, writer, and associate member of the ASPR who was also a friend of Hodgson's. About a month after Pellew's death, Hodgson had a sitting with Piper and was startled by Pellew's arrival—even more so when the spirit revealed all sorts of private details about Hodgson. While Hodgson might believe that Piper's

G. Stanley Hall (seated, center) with Sigmund Freud to his right and Carl Jung to his left during the European pair's visit to the USA in 1909.

control was his friend, other researchers—not to mention Pellew's own family—were less convinced. Scottish folklorist Andrew Lang noted that Pellew seemed to have forgotten much that he'd known while alive, such as philosophy, classical Greek, and even details of his own biography. It's difficult not to conclude that Piper—who had by now been working with the obsessed Hodgson for five years or so—gained those secrets via nonpsychic means.

Hodgson became so convinced by Piper that his faith in her survived even her abrupt recantation in 1901, when she renounced mediumship forever and admitted she probably wasn't in contact with spirits. (In this case, "forever" didn't last long; she was soon back in the

⇒ LESLIE FLINT ⇐

One of the best-known mediums of the twentieth century was the Londoner Leslie Flint. He was a direct-voice medium, meaning that the spirits supposedly speaking through him did so in their own voices. This is obviously a much harder trick for the medium to pull off than most, because it's all too easy not to know that the sitter's Auntie Hilda had a thick Irish accent. Speaking of Irish accents, one of the many famous people whose spirits spoke through Flint was George Bernard Shaw, and he had somehow *lost* his! Typical of the many challenges to Flint was that of one Gladys Lorrimore, who complained that the spirit voice of her late husband sounded nothing at all like the one he'd had when he was alive.

The same didn't hold true, however, for Eira Conacher, who was so convinced that Flint had put her in touch with her late husband, Douglas, that she recorded the sittings and, with Flint's acquiescence, published the results in two books, *Chapters of Experience* (1973) and *There Is Life after Death* (1978). In my young and impressionable days, I was Eira Conacher's editor for *Chapters of Experience*, and accompanied her to a few of the media events that the publicists set up. Eira managed to persuade Les to participate in a live interview on BBC Radio. Naturally, the interviewer asked Les to try to get in contact with the spirits on air; equally naturally, Les couldn't—because sometimes, you know, they're just not in the mood.

It was at that moment that, watching from the control box, I realized there might be something rotten at the heart of the book I'd helped publish. Since I very much liked Eira Conacher, I found this quite worrying for a while. And then I realized that, cheeringly, while Les had essentially written—well, dictated—the book *Chapters of Experience*, Eira was the one getting all the royalties!

fold.) To add to his indignities, soon after his death in 1905, he became Piper's latest control.[10] In 1909, at one of the sittings where "Hodgson" was speaking through the medium, the psychologist G. Stanley Hall requested the presence of his dead niece, Bessie Beals. Sure enough, Bessie could be found on the Other Side. Overjoyed to be in contact with her uncle, she gushed appropriately about her memories of him. Unfortunately, as Hall eventually revealed, he'd never had a niece called "Bessie." As smoothly as you like, "Hodgson" disclosed that—oops, easy enough mistake to make—the spirit was not a Bessie but a "Jessie" who had been a loved one of a different sitter present at the séance. Another time "Hodgson" was asked to name six of his buddies at Boston's Tavern Club. The six names the spirit gave included five men who weren't his buddies and a sixth who wasn't a member of the club.

It intrigued William James that Hodgson was now one of Piper's controls. Although James had, through pressure of other work, given up investigating Piper years before, he now reentered her life. Over the next three years he worked on the Hodgson communications, eventually concluding that there was no real evidence of anything supernatural at work.

❧ MISS X

A notorious instance of senior officials of the SPR being hood-winked concerned F.W.H. Myers and the aristocratic medium Ada Goodrich-Freer, whom Myers met (and was probably seduced by) in the late 1880s. Myers was a womanizer, but Goodrich-Freer must have been something out of the ordinary because she turned his head so far that tricking him became easier for her than stealing candy from a baby. Right at the outset, she deceived him—and many others within the SPR—as to her antecedents: She said she

---

[10] F.W.H. Myers had already become another.

was a descendant of a noble Scottish family; she was, in fact, the daughter of a Leicestershire vet. Neither was she, as she asserted, the first female fellow of the Royal Society. And she was not a teenager when Myers met her, as she told everyone, but a thirty-year-old woman, having been born in 1857.

Ordinarily, this was the kind of stuff that Myers would have checked out, but in the case of Goodrich-Freer he didn't. When, in the 1890s, the SPR decided to carry out an investigation into the phenomenon of second sight among Scottish Highlanders, Myers was in the forefront of those recommending Goodrich-Freer, with her "distinguished Scots lineage," for the job. Off she went northward, soon to return with an abundance of first-rate corroborative material—all of which she'd plagiarized from the Scottish folklorist Father Allan McDonald.

Clandon House, a stately pile that the notorious Ada Goodrich-Freer claimed was haunted.

In 1895, Goodrich-Freer investigated the stately home Clandon Park (near Guildford, England) on suspicion that it was haunted. At least one of her fellow guests reported that Goodrich-Freer had told him she'd seen nothing during the night she spent there, yet later she wrote in the SPR's *Journal* that she and others had seen a hooded female ghost. Clandon Park's owner, Lord Onslow, who derived income from renting the place out to tenants, took vigorous exception to this fantasy, and Goodrich-Freer—and the SPR—had to back down.

Ballechin House (near Dunkeld, Scotland) was her next puta-tively ghost-infested stately pile. Some tenants had been frightened off by local rumors that the place was haunted by the ghost of a soldier named Robert Steuart and several of his dogs, and in 1897 Goodrich-Freer persuaded John Crichton-Stuart, third marquess of Bute—a fellow researcher of the paranormal—to rent the place for her so she could check it out. Lord Onslow's righteous wrath as land-lord of Clandon Park had taught her a lesson; Bute rented the place in the name of a certain Colonel Taylor, who claimed he wanted to go there for a bit of huntin' and fishin'. During Goodrich-Freer's tenancy, some forty guests were invited for shorter periods. Curiously, few of them reported much spectral activity, yet Goodrich-Freer somehow detected a plethora of poltergeist and other ghostly activity—often occurring in rooms whose occupants had sensed nothing! More than one of these guests began to suspect that any odd sounds or sights they might experience were being engineered by Goodrich-Freer herself. John Milne, the geologist and seismogra-pher, had an alternative explanation: Ballechin House might be built on or close to a patch of geologically unstable terrain, so that the bumps and crashes were being produced not by spirits but by minor earth tremors. (We'll encounter similar theories later; see page 96.) Even so, enough people were convinced the place was haunted that Goodrich-Freer returned to London confident that she had the material for a book, *The Alleged Haunting of Ballechin House* (1899).

Unfortunately for her, while she was still writing the book, a number of the short-term guests went public with their suspicions of skullduggery. Myers, whose roving eye had by now roved onward a few times, declined to support her side of the argument, and she was disowned by the Society. Even so, she went ahead and published. Goodrich-Freer, who sometimes used the pseudonym "Miss X," persevered as a psychic in London until she was caught cheating at a séance. She eventually emigrated to Jerusalem, where she married an American scholar sixteen years her junior whom she easily persuaded was two years her senior!

✳ AUTOMATIC WRITING AND THE OUIJA BOARD

In automatic writing, the subject (known as an automatist) rests a hand, holding a pen or pencil, on a sheet of paper and lets the mind go blank—or, perhaps, goes into a trance. Sooner or later the "spirits take control," and the hand starts to move, writing the messages they dictate from the Great Beyond . . .

One possible explanation for automatic writing lies in the way that the brain makes decisions to take action. Experiments have consistently shown that, when the brain decides on something, about one-third of a second passes before it informs the conscious *you* that it has done so. Of course, you (your conscious mind) think you've made the decision yourself—which in a very definite sense you have, since you've assembled and weighed all

G. W. Cottrell launched his version of the planchette in Boston in 1860.

⇛ CHANNELING THE MASTERS ⇚

The UK's Matthew Manning practices now as a healer (and has written a number of books on the subject), but his voyage of paranormal discovery began during his childhood, when his family home was the center of outbreaks of poltergeist activity. This eased off a bit when he discovered he could do automatic writing. Apparently, when he was taking dictation from the spirits, they had less inclination to throw the furniture around. (A cynic might say that the furniture wasn't thrown around as much when Matthew's hands were occupied in automatic writing.) The first season of poltergeist mayhem occurred when he was eleven, studying for his Common Entrance examination; there was a recurrence a few years later, this time seemingly linked to his O-Level examinations.[1] The poltergeist outbursts were investigated by the parapsychologist George Owen, who believed they were genuine.

By the time Manning was eighteen, Owen[2] was working in Toronto, directing parapsychological studies at the New Horizons Research Association. Owen brought the youth across to Canada to see if he could replicate the Israeli illusionist Uri Geller's cutlery-bending feats in a controlled setting. Manning showed in the laboratory that he was capable of such deeds. More interesting, according to Renée Haynes in her history of the SPR, was Owen's discovery that each act of cutlery-bending was preceded by a shift in Manning's brain waves, as recorded by an electroencephalograph. I'm not certain why this should be regarded as in any way remarkable; just before we consciously decide to do something, the patterns of our brain waves typically change. In effect, all the EEG showed was that Manning had made a decision. It didn't show what he had decided to do.

At some stage, Manning added a new ability to his repertoire: As well as automatic writing, he discovered he could now do automatic art, somehow channeling the talent of great dead artists such as Picasso, Beardsley, Rowlandson,

Bewick, and Matisse. In fact, as he told the *San Francisco Examiner*, one of his Picassos was so similar to an original that appraisers at the London auction house Sotheby's had told him they would have accepted it as such had he not informed them otherwise.

Similarly, as Manning later reported to a UK newspaper, the *Daily Mirror*, a piece of artwork that he had reproduced in *The Link: The Extraordinary Gifts of a Teenage Psychic* (1974) had caused a commotion at the Rijksmuseum because it was identical to a Jan Savery painting in their vaults, which had never been published.

In the former instance, James Randi, the professional conjurer and world-famous debunker of matters

Jan Savery's painting of a dodo, 1651.

woo, reported that he wrote to the Sotheby's appraisers and got the reply that they had, indeed, assessed Manning's pictures and judged them obvious fakes. Randi also reported that he wrote to the Rijksmuseum and was informed that the Savery painting was on public display and that it had been reproduced in print at least twice.

---

[1] Both were features of the UK school system at the time; they've long since been replaced.

[2] We'll meet Owen again in connection with "Philip" on page 105.

the factors that determined the decision. But the actual decision itself is made in one part of the brain and, at the same time that appropriate signals are being sent to the muscles to take action, signals are also being sent to a different part of the brain, which both informs you that the decision has been made and creates the illusion that it was *you* who made it.

This has some consequences for philosophical concepts of free will that we'll hastily skim over. In the context of automatic writing, however, what may be happening is that the last part of the process—the sending of messages to you to tell you what's going on—is temporarily malfunctioning or suppressed. Thus, the brain is deciding to write all kinds of stuff while the conscious you is unaware that your own brain made those decisions.

Supposedly invented in France in 1853, the planchette swiftly became popular among amateur Spiritualists as a facilitator of automatic writing. The device—essentially a heart-shaped board on wheels, with a hole drilled through it to hold a downward-pointing pencil—is put on a sheet of paper; the user rests a hand on the device. This setup supposedly makes it easier for the spirits to move the automatist's hand around.

The novelty maker Elijah Bond patented the Ouija board in 1890. The name *ouija* is popularly believed to be a combination of the French and German words for *yes*, but Baltimore businessman Charles Kennard claimed to have derived the name from the Egyptian word for "good luck."[11] This retained the planchette but omitted the pencil and paper; the spirit messages were now supposedly conveyed by the planchette pointing to the letters of an alphabet printed on a wooden board. Although this was a far more laborious means of receiving messages, it caught on in a way the planchette never had—perhaps because the

---

[11] Some sources contend that it was first named as such in 1901 by the toymaker William Fuld, who had taken over the manufacture of the boards.

Ouija board, marketed as a family game,[12] lent itself to group participation: Several people could rest their fingers on the pointer as it spookily moved around the board.

A long series of experiments into how the planchette and the Ouija board might function was carried out in the 1890s by the American psychologist Joseph Jastrow. Rather in the same way that Michael Faraday had earlier created a special table that could record people's unconscious actions (page 10), Jastrow created a planchette that was especially sensitive to people's hand movements. It comprised two sheets of glass, one above the other,

The psychologist Joseph Jastrow, one of the great debunkers of fake claims about the psychic.

separated by metal balls. As there was very little friction, even the tiniest impulse would move the upper plate relative to the lower. A stylus and a soot-blackened piece of paper—both hidden from the subject's view beneath the upper sheet of glass—completed the device. By talking about objects in the room or places outside it, Jastrow induced his subjects to make unconscious hand movements that were duly recorded by the stylus scratching the sooty paper.

Of course, there's a distance to be traveled between small unconscious hand movements and the larger-scale motion of the Ouija board's planchette as it goes from letter to letter. The likely cause of such large-scale motion is the reinforcement of one person's ideomotor responses—responses like the hand movements Jastrow measured—by the cooperation of the other participants. Just as we have a tendency to agree with other people, even when we don't

---

[12] Ironically, it's occasionally attacked by the religious as an invention of the Devil, as it's all too easy for Satan, working through the Ouija board, to possess the mind of the user.

particularly support their argument, so we tend to respond to someone's small finger pressure on the planchette by going along with it. With several people's fingers all providing a small impetus in one direction, the planchette starts moving more resolutely . . .

The humble Ouija board has made its mark on literature in various ways. The American medium Jane Roberts was allegedly first contacted by a spirit entity named Seth via a Ouija board, although she was later able to channel Seth directly. (The renowned "Seth Material," a couple dozen books supposedly dictated by Seth through Roberts to her husband, have sold millions of copies.) In some ways even more remarkable is the epic poem *The Changing Light at Sandover* by Pulitzer Prize–winning poet James Merrill, which appeared in three volumes between 1976 and 1980. Large parts of this poem were derived by Merrill from Ouija-board sessions conducted by himself and his partner, David Jackson. Many of Merrill's other works were written by the same means.

PATIENCE WORTH

Pearl Curran was a young Missouri housewife of no special note until July 1913, when she and her friend Emily Hutchings, playing with a Ouija board, started to receive communications from a spirit who described herself as Patience Worth, a woman born in 1649 (or 1694; she was inconsistent about this) in the county of Dorset in the English Southwest who, in adulthood, had come to North America and finally met her end at the hands of vengeful Indians. For the rest of Curran's life—she died in 1937—she would "take dictation" from Patience to the tune of nearly four million words. After the publication of *Patience Worth: A Psychic Mystery* (1916) by the journalist Casper S. Yost, Curran and Patience entered the limelight.[13] There

---

[13] A much more rigorous study by psychic researcher Walter Franklin Prince, *The Case of Patience Worth*, came along in 1927.

was a surge in the sale of Ouija boards (and also a surge in the condemnation of said boards by Christian authorities).

Moreover, Patience proved herself to be quite a writer, publishing numerous poems, a play, and several novels. And these weren't drivel. In 1918 the Joint Committee of Literary Arts of New York hailed her as one of the year's outstanding authors. The *New York Globe*'s reviewer claimed her novel *The Sorry Tale* was as good as Lew Wallace's *Ben-Hur* (1880). Her poems appeared in prestigious anthologies and magazines, alongside poems by the likes of Edna St. Vincent Millay. Novels like *The Sorry Tale: A Story of the Time of Christ* (1917), *Hope Trueblood* (1918), and *The Pot upon the Wheel* (1921) found a ready audience, as did *Light from Beyond* (1923), a collection of the best of Patience's poetry. The richness of Patience's language and its quirky use of (supposed) Ozark expressions and syntax impressed many. The fact that *Hope Trueblood* was set in Victorian England—a couple of centuries after Patience's supposed death—raised doubts, as did the fact that the text was written in an entirely different style.

Both at the time and more recently, psychologists have tended to conclude that "Patience Worth" was either the unconscious or a secondary personality of Curran. According to the ASPR's James Hyslop, writing in 1916, she might simply have been a fraud: Pearl Curran and her husband James were far from illiterate and, between them, had enough knowledge that they could easily have concocted the texts. An additional problem was that, outside Patience's own statements, evidence of her seventeenth-century existence stubbornly refused to emerge. Although of course in those times not a great deal of attention tended to be paid to women of humble birth, Patience should have turned up in *some* record or another.

A hint at the truth may lie in "Rosa Alvaro, Entrante," a short story that Pearl Curran wrote under her own name and which was published in the *Saturday Evening Post* in 1919. After a bogus psychic tells a lonely young woman, Mayme, that she has a flamboyant Spanish spirit guide, Rosa Alvaro, Mayme enjoys slipping from one persona to the other before eventually confessing to a friend that she knows Rosa doesn't really exist—that Mayme just enjoys the freedom to behave as Rosa would rather than be confined to her own inhibited self. The story was made by the Goldwyn Company into the movie *What Happened to Rosa* (1920), starring silent film comedienne Mabel Normand.

Mabel Normand starred as the flamboyant spirit guide in *What Happened to Rosa* (1920).

The friend with whom Pearl Curran first started to play with a Ouija board, Emily Hutchings, started channeling a dead author of her own. This time, though, the author

Did Mark Twain write a final novel, *Jap Herron* (1917), from beyond the grave? Not if *he* had anything to say about it!

was not some seventeenth-century nonentity but the great Mark Twain—an odd choice, because Twain was famously skeptical about all matters psychic. In 1917 "Twain's" *Jap Herron: A Novel Written from the Ouija Board* appeared, only to become the focus of a major legal case when Harper & Brothers, who owned the rights to Twain's works, claimed that the critically panned *Jap Herron* could damage sales of the Twain books they still had in print. Before the case came to trial, Hutchings's publisher settled out of court and withdrew the novel from sale.

❧ CONNING DOYLE

In marked contrast to Twain, there was the great champion of Spiritualism (and much crackpottery) during the early decades of the last century: the writer Sir Arthur Conan Doyle, creator of such literary icons as Sherlock Holmes and Professor Challenger. In *Flim-Flam!* (1980), James Randi points out quite forcefully the reason why charlatans could so easily pull the wool over Doyle's eyes. Randi observes that the

Conan Doyle's *alter ego* Sherlock Holmes (right, as played by William Gillette in a 1900 New York production) injects himself with a cocaine solution as Watson looks on with dread.

master detective Holmes would not in fact have been able to function outside the artificial world Doyle created for him:

> For his deductions to be correct, the consistency of his world was absolutely necessary. People in particular had to conform to type; otherwise Holmes would have been hopelessly wrong [in his deductions]. It was just this rather naively invented universe that Doyle imagined into existence and projected about himself, and it accounts in large measure for his fanciful interpretation of phenomena that he came upon only late in life—the wonders of spiritualism.

Doyle thus expected people to tell the truth unless they were "obviously" liars. He was also an irremediable egotist, having an enormous

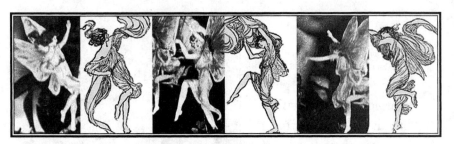

It was pointed out that the "fairies" were remarkably like some of Claude Arthur Shepperson's illustrations in *Princess Mary's Gift Book* (1914).

and quite unfounded faith in himself as a mighty thinker—since that was how the public largely regarded him, having confused the fictional creation with the man. Like many a movie star in our own age, enormous popular success, widespread public adulation, a circle of sycophants, and the possession of considerable wealth had all conspired to give him a false sense of his own intellectual ability. Once he'd decided upon something, there were few who dared contradict him. And thus he most famously fell hook, line, and sinker for the obviously faked photos two young girls, Frances Griffiths and Elsie Wright, had supposedly taken in 1917 of fairies in Cottingley Glen, near Bradford, Yorkshire.

Four years later the two girls were brought back to the glen and given a stereoscopic camera and a cine camera. While everyone hung around expectantly, the only thing that materialized was a whole lot of rain. Perhaps the fairies were reluctant to come out in the rain? Perhaps. But Doyle had an alternative explanation: "A small seam of coal had been found in the Fairy Glen and it had been greatly polluted by human magnetism."[14] Well, you can hardly expect to see troops of fairies in circumstances like that, can you? And there was more: "These conditions might perhaps have been overcome; but the chief impediment of all was the change in the girls, the one through womanhood and the

---

[14] Doyle, *The Coming of the Fairies* (1922).

other through board-school education." Of course, we all knew that allowing women to get an education was going to lead to disasters like losing the ability to see fairies at the bottom of gardens.

The Cottingley fairies pictures weren't the only doctored photographs that Doyle fell for. From its formation in 1919, he was a member of the Society for the Study of Supernormal Pictures and led a mass resignation from the SPR after psychic researcher Harry Price had exposed the fraudulence of the "spirit photographer" William Hope (page 115). He also wrote the book *The Case for Spirit Photography* (1922).

⇒ VALIANTINE'S TOEPRINT ⇐

During a séance, the direct-voice medium George Valiantine once produced what he claimed was the thumbprint of Sir Arthur Conan Doyle. It proved to be a print of Valiantine's own big toe. Repeatedly exposed as a faker by investigators from the SPR, Valiantine was also one of those tested by the investigating committee for the 1923 *Scientific American* award (page 30). Unknown to Valiantine, who impressed the investigators by his ability to attract spirits that talked with each other high in the air above the sitters' heads, on the third of the three trials his chair was rigged so that, should he leave it, a light would go on in a different room. Sure enough, the light went on—several times.

Bizarrely, at least one of the investigators, the writer H. Dennis Bradley, insisted vehemently this was not actual cast-iron proof that Valiantine had been cheating. (Later in life, after years of vigorously supporting Valiantine, Bradley reversed his opinion entirely.)

After Doyle's death in 1930, mediums on both side of the Atlantic were quick to claim visitations by his spirit. (Who better to confirm to their sitters that, yes, the afterlife was just as they imagined it?) In the end Doyle's widow, Jean, became so irked by the impostures—seemingly because she felt that all the bogus Doyles were drowning out any real manifestations her late husband might be making—that she began issuing veiled threats of legal action. These had no effect whatsoever.

❧ MAGIC MACHINATIONS

James Randi is not the first conjurer to have put his prestidigitatory skills toward the pursuit of fakery among psychics and paranormalists. Harry Houdini famously did the same thing. Initially a good friend of Arthur Conan Doyle, Houdini grew ever more to abhor fraudulent Spiritualists and Spiritualism in general. Somehow their friendship survived despite Doyle's acceptance of virtually all matters occult—until the final straw came.

The end began in June 1922, when Doyle's wife, Jean—who fancied herself as an automatist—received (so she said) a message from Houdini's dearly beloved mother. Written in English, the screed that Jean's hand wrote was preceded by a devout cross. Since Houdini's mother had barely been able to speak English and was unable to write in the language (while she

A faked spirit photograph of Harry Houdini with the ghost of Abraham Lincoln, c. 1925.

was alive, she always wrote to her son in German), and since she had been Jewish—not Christian—Houdini could see immediately that the message was bogus.

Although he was polite about it, Houdini's natural conclusion was that Lady Doyle had forged the "spiritual communication" deliberately in an attempt to emotionally coerce him into agreement with her husband. Even so, he played along when, immediately after reading the message, the Doyles asked him to open his mind to receiving messages from the spirits directly. Houdini wrote down the name "Powell," and Doyle was thunderstruck. His friend, Ellis Powell, had recently died. In vain, Houdini tried to convince Doyle that he'd been thinking of his own friend and fellow-conjurer Frederick Eugene Powell, who was currently experiencing difficulties. From then on it was Doyle's fixed belief that Houdini was secretly a psychic who was using his powers both to quash the efforts of the mediums he was supposedly exposing and, of course, to perform his stage tricks.

In June 2015 *Variety* reported on a new TV show, *Houdini & Doyle*, due in 2016, in which Houdini (Michael Weston) and Doyle (Stephen Mangan) solve mysteries for Scotland Yard. Houdini must be spinning in his grave.

Away from the séance room, Houdini also exposed people purporting to possess powers of what we'd today call ESP—including the giant teenager, Joaquín María Argamasilla. Known as "the Spaniard with the X-ray eyes," Argamasilla fooled not just lay audiences but also scientists like Gustav Geley and Charles Richet with his claims of being able to see through metal. His two principal tricks were reading the randomly set

Doyle, shown in this 1914 photo, came to believe that Houdini was really a powerful psychic and just *pretending* to be a conjurer.

time on a pocket watch through its case and reading cards, written messages, or the numbers on dice placed in locked metal boxes. Houdini witnessed a demonstration by the youth in 1924 and was immediately able to detect the childishly simple methods used to pull off the illusions. In the case of the watch trick, Argamasilla simply distracted the spectators long enough to open the case for a quick peek!

Also in 1924 came Houdini's arguably greatest challenge, at least insofar as exposing fraudulent psychics was concerned: the Boston medium Mina Crandon,

Dating from about 1909, a poster for one of Houdini's debunking presentations.

also known as "Margery." Mina had been introduced to mediumship only in the previous year by her much older husband, Dr. Le Roi Goddard Crandon. Soon she was displaying a wide range of mediumistic skills, and, after she had impressed many of the locals, Crandon took her to Europe, where she similarly impressed the likes of Richet and Doyle. During her séances she produced apports, made tables dance using a "psychic structure" (a sort of ectoplasmic limb that supposedly emerged from between her legs), and all the rest of it. Her spirit guide was Walter Stinson, a brother of hers who had died in a train crash years before. Walter spoke in a coarse, aggressive fashion, his language peppered with manly expletives, and it seems to have been this saltiness that persuaded witnesses more than anything else that the sweet young Mina herself could not possibly be responsible for the voice.

Back in the United States, Mina was a natural to compete for the prize being offered by *Scientific American* to any psychic who could persuade its committee that his or her claims were valid. Although the terms of the contest were that the psychics come to New York (all expenses paid) to give their demonstrations, Dr. Crandon counteroffered to pay the costs of any relevant *Scientific American* staff to come to Boston and stay as guests in his home during their investigation, making the excuse that he couldn't leave his medical practice. This should perhaps have raised the suspicions of J. Malcolm Bird, the magazine's associate editor, who was acting as secretary to the committee—after all, surely Mina could have traveled to New York alone or with a friend—but it didn't. The Harvard psychology professor William McDougall attended one of her séances for a preliminary check, and further investigation was carried out by Bird and another committee member, the parapsychologist Hereward Carrington, both of whom stayed as guests of the Crandons several times.

The first Houdini heard of the investigation seems to have been at the insistence of *Scientific American*'s publisher, Orson Munn, when Bird and Carrington were apparently well on their way to declaring Mina a valid medium—that is, to *Scientific American*'s committee, and thereby the magazine itself, accepting that Spiritualist phenomena were real. Houdini and Munn rushed to Boston and, declining Crandon's offer to put them up, stayed at a local hotel. Houdini was appalled that Bird and Carrington had compromised their objectivity by letting the Crandons be their hosts, and for the first séance he attended, he made an odd preparation: He tightly bandaged his lower leg for some hours and then, not long before the séance, tore the bandages off so that the skin of his lower legs was highly sensitive. As part of her standard performance, Mina rang a little bell that was in a box placed between the feet of the person sitting next to her—the reasoning being that she could hardly reach the box with her own feet if someone's leg was in the way—so Houdini arranged to be that sitter. Sure enough, sitting in the darkness, his leg bared and sensitive, he was perfectly able to feel Mina sneak her foot around to ring the bell.

By the end of that séance, Houdini had worked out how Mina performed all her tricks. Bird and Carrington put up a ferocious resistance to his analysis, and the upshot was a challenge being issued that Mina should perform a further séance—this time under controls devised by Houdini himself. He built a wooden-box arrangement that he reckoned made it impossible for Mina to cheat, at least physically. When the Crandons saw it, they were horrified but had little option other than to cooperate. According to Crandon (cited in Milbourne Christopher's 1969 biography *Houdini: The Untold Story*):

> *The psychic does not refuse to sit in the cage made by Houdini for the committee: but she makes the reservation that she knows no precedent in psychic research where a medium has been so enclosed; and she believes that such a close cage gives little or no regard for the theory and experience of the psychic structure or mechanism.*

In other words, while there were holes for her hands and feet, Houdini had deliberately omitted to include an aperture for that ectoplasmic limb!

The Crandons insisted on holding a practice séance with friends before the main encounter in the presence of the committee. Then, the first time the committee members sat in on the proceedings, Mina was able to summon Walter, as usual, but do little else except break open the front of the box, which was rather flimsily constructed. (Crandon insisted that it was Walter who'd pushed apart the construction.) The second time, nothing happened until eventually Walter asked Daniel Fisk Comstock, another committee member, to examine the bell-box under a light. Comstock did so, and discovered that its flap had been jammed using a little pencil eraser.

At a later séance, Walter accused Houdini of attempting to rig matters so that Mina couldn't possibly enact her spiritual phenomena—psychic structure or no psychic structure! It's fairly obvious, however, who did the rigging (just as it seems fairly obvious who supplied the voice of "Walter"). Once put in a position where they couldn't perform any trickery, the Crandons figured that the best plan was to make it seem as though Houdini were the cheater. Many people believed (and continue to believe) the version of events the "spirit" offered, even though Houdini published an illustrated pamphlet detailing each of Mina's fraudulent methods.

Scientific American declined to validate Mina Crandon's mediumship and published an article explaining why. But matters were complicated by the fact that the ASPR—which had recently been taken over by a more pro-Spiritualism faction—backed her. The ASPR's chief research officer, Walter Franklin Prince, who fervently believed Mrs. Crandon was a fraud, then departed to serve the rival Boston Society for Psychical Research, and the controversy rumbled on for a number of years.

It's no wonder so many psychic investigators fell under Mina's spell. She reportedly performed some séances in the nude and made a point of throwing herself onto the laps of the male sitters. Sometimes she would even sprinkle luminous powder on her breasts.

When, in later years, Hereward Carrington admitted to friends that he'd been having an affair with Mina all along, he hastened to reassure them that this in no way influenced his judgment. Yes, right.

❧ EROTIC ECTOPLASM

It can't be denied that one of the great attractions of the séance was the opportunity it gave men—sorry, objective investigators of knowledge's frontiers—to mingle more intimately with attractive young women than the strictures of society would normally allow. As time went on, the sexuality involved in séances grew ever more overt. The growing fad for ectoplasm hastened the process. If ectoplasm were supposedly emanating from every possible orifice of the medium, was it not the duty of the investigator to examine said orifices closely? Of all the mediums whose exploitation of sexuality has been recorded, probably the most famous, aside from Eusapia Palladino, was

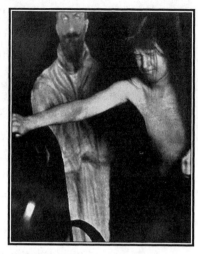

The medium Eva C. with a "spirit presence" that looks alarmingly like a cardboard cutout.

"Eva Carrière" or just "Eva C." (born Marthe Béraud). Her specialty was materializing human faces in the midst of ectoplasm. Although she apparently admitted to friends in private that her act was just that—an act—it seems to have been an accomplished one.

One man who did not rush to judgment while investigating her was German physician Baron Albert von Schrenck-Notzing, who published the results of his extensive ectoplasmic researches in *Phenomena of Materialization: A Contribution to the Investigation of Mediumistic Teleplastics* (1920).[15] The description[16] of Eva C.'s September 10, 1911, séance became quite notorious:

> *A large, flat, dark-grey patch appeared on her breast, white at the rims. It remained for some time, and then disappeared in the region of the navel. I clearly saw it being reabsorbed there.*

---

[15] "WITH 225 ILLUSTRATIONS," trumpets the title page alluringly.

[16] Written in this instance by Eva's companion, Juliette Bisson.

The curtains were then kept closed for several seconds, without my releasing her hands. A round patch again appeared on her skin at the opening of the curtains. It had the same kind of shape as the first, but was larger. To this was joined, in the left ovarial region, a large, black, ball-shaped structure, white in the middle and dark grey at the rims. With the curtain open, I counted twenty-two seconds. Suddenly the material folded itself together at right angles to the axis of her body, and formed a broad band extending from hip to hip under the navel. This apparition then folded up and disappeared in the vagina.

On my expressing a wish, the medium parted her thighs and I saw that the material assumed a curious shape, resembling an orchid, decreased slowly, and entered the vagina. During the whole process I held her hands. Eva then said, "Wait, we will try to facilitate the passage." She rose, mounted on the chair, and sat down on one of the armrests, her feet touching the seat. Before my eyes, and with the curtain open, a large spherical mass, about 8 inches [20 cm] in diameter, emerged from the vagina and quickly placed itself on her left thigh while she crossed her legs. I distinctly recognized in the mass a still unfinished face, whose eyes looked at me. As I bent forward in order to see better, this head-like structure rose before my eyes, and suddenly vanished into the dark of the cabinet away from the medium, disappearing from my view. Again the medium fainted.

When Eva, in her armchair, had recovered her somnambulic consciousness, I saw, eight times in succession, a head covered with veils, which was quite detached from the medium. The phenomena then ceased, and I closed the sitting at 10 p.m.

It was Eva C.'s habit to allow her sitters to inspect her vagina both before and after a materialization séance to make sure she wasn't simply stowing props there. As with Eusapia Palladino's orgasmic moans, this must surely have compromised many an investigator's objectivity on the matter.

Predictably, Arthur Conan Doyle found Eva C.'s performance utterly convincing. However, Harry Houdini was less impressed, recalling in *A Magician among the Spirits* (1924) a séance of hers that he had witnessed with associates:

> I saw distinctly that [the supposed ectoplasm] was a heavy froth and was adhering to her veil on the inside. [It] had emanated from her mouth, but when she leaned forward it looked as though it was coming from her nose. She produced a white plaster and eventually managed to juggle it over her eye. There was a face in it which looked to me like a colored cartoon and seemed to have been unrolled.

For a long time, there was no grand moment of exposure of Eva C. as there was—several times over—with Palladino. Rather, as the years went by, more and more people became convinced that she was a fraud for the same sort of reasons as Houdini had: The ectoplasm and the materializing faces somehow just didn't look quite right. But then, in 1914, close examination by a certain Miss Barkley of a photo Schrenck-Notzing had taken of a materialized face from above and behind revealed the letters "LE MIRO." From that point it didn't take long to demonstrate that the faces were not so much material-ized as mundanely material—Eva had cut pictures out of *Le Miroir* and other popular magazines and then, with a pen, had given them beards, wrinkles, and so on in hopes of disguising them.

Miss Barkley published her research and the public roared with laughter. Schrenck-Notzing tried to fight back, accepting that the faces did indeed resemble the magazine photos but claiming that they were "ideoplasts"—images born from Eva's mind (page 70). Was it not likely that Eva had seen the magazines in question and that the memories of the images were hidden in her mind, being accessed unwittingly in the form of materialized faces during her séances? The argument convinced few.

For a number of years, Eva lived with a widow named Juliette Bisson. It was during this period that Houdini witnessed her in séance, and afterward he declared both that Bisson was an accomplice and that the two women were lovers; later, he was proven correct on both counts. Eventually, in 1920, Eva married and her mediumship abilities vanished.

HELEN DUNCAN

One of the many reasons the investigators of spirit mediums have suspicious minds is the case of Helen Duncan. In the 1920s, 1930s, and 1940s, this Scottish medium did a good trade in séances, especially captivating sitters through her production of ectoplasm and spirit forms. She made the mistake in 1928 of permitting the photographer Harvey Metcalfe to take a few pictures using flash during one of her séances, and later examinations of these showed that the spirits were rendering themselves in the distinctly nonethereal form of butter muslin. In 1931, the psychic researcher Harry Price—of Borley Rectory fame (page 86)—repeated the photographic experiment with similar results, identifying butter muslin, a rubber glove, toilet paper, photos cut out from magazines, a papier-mâché mask, safety pins . . .

A 1931 photo by psychic investigator Harry Price of the medium Helen Duncan regurgitating a spirit formed from either ectoplasm or . . . cheesecloth?

➤ ARCANE TISSUE? INVISIBLE ECTOPLASM ➤

Charles Richet, who coined the term *ectoplasm*, thought it was an expression of *ideoplasty* (another term he coined): If the human mind can produce physiological changes such as stigmata or psychosomatic blisters, why can't it make the body exude a form of tissue hitherto unknown to medical science? The speculation seems an odd one, coming from a physiologist who must surely have known there was exactly zero physical evidence that the body—whatever the mind might instruct it to do—is capable of producing masses of arcane tissue.

Oliver Lodge, who with Richet investigated Eusapia Palladino and concluded that she was genuine, posited that she was able to shift objects that were beyond her reach through her ability to produce invisible ectoplasm, such as "an invisible ectoplasmic rod—that is, some structure unknown to science, which could transmit force to a distance" (*Past Years: An Autobiography*, 1931).

A few years later she was tested by the London Spiritualist Alliance, which established that one of her tricks was to swallow her "ectoplasm" and then puke it up on demand. And, at an Edinburgh séance in 1933, one of the sitters grabbed Duncan's spirit guide—a little girl called "Peggy"—and discovered that the "spirit" was in fact a ladies' undergarment. Duncan was arrested for fraud and fined £10.

Her more serious brush with the law came in 1944, when the Portsmouth, England, police raided one of her séances. The concern of the authorities was less that she was a phony medium and more that she was using dishonestly obtained military intelligence to boost her claims to be bearing communications from the spirit realm. The case eventually reached the Old Bailey, where Duncan found herself facing several charges, including (among others) conspiracy to contravene the Witchcraft Act of 1735 (the prosecution was obviously in dire straits if it had to dig this dusty statute out of history's wastebin). Found guilty of this charge, she had the dubious honor of being the last person ever convicted under the Witchcraft Act and was sentenced to nine months in prison.

A long-standing myth is that Helen Duncan's death in 1956 was due to police raiding a séance she was conducting: When they switched the lights on, apparently, the ectoplasm she'd been producing snapped back into her body at such speed that it killed her. Some mediums have indeed maintained that ectoplasm is connected to the nervous system, so that a sudden switching on of lights and retraction of ectoplasm might be dangerous. Since the mediums knew very well where their ectoplasm came from, it seems obvious that this was merely a way of inhibiting sitters from switching on the lights.

Helen Duncan's death was actually due to heart failure. She was grossly obese, so it was bound to happen sooner rather than later.

DORIS STOKES

One medium who was inspired by attending a séance of Helen Duncan's was Doris Stokes. In the 1970s and 1980s, Stokes became arguably the most famous Spiritualist medium of the modern era—certainly in her native UK, where she was the subject of numerous television broadcasts, was regularly in the headlines of the tabloid newspapers, and put on mediumistic performances in the largest theaters in front of sell-out crowds. In his book *The After Death Experience* (1987), Ian Wilson describes a show she put on in 1986, near the end of her life, at one of the United Kingdom's biggest venues, the London Palladium, which was attended not only by Wilson himself but by a pair of TV journalists. The journalists had intended merely to obtain contact details, for the purpose of follow-up interviews, from some of the audience members for whom Stokes had produced dramatically accurate readings. Instead, they discovered that those apparently randomly chosen members of the audience had all been invited and given free tickets by Stokes, on the basis of information given to her by family members who thought they might benefit from being put in touch with their deceased loved ones. So all the names and circumstantial details that Stokes had apparently gained from the Other Side had been available to her beforehand through perfectly mundane channels.

Stokes also tried her hand at psychic detection, intervening with an extravagant lack of success in the case of the Yorkshire Ripper—the serial killer Peter Sutcliffe, who murdered thirteen young women between 1975 and 1980. In one of her autobiographies, *Voices in My Ear* (1980), Stokes claimed to have helped solve two murder cases in her native Lancashire, to the joy and gratitude of the local police. Unfortunately, the Lancashire Constabulary has no recollection of this. She also later claimed that during a 1982 US tour she assisted the Baltimore County Police in the case of missing teenager Jamie Griffin. Colonel Joseph A. Shaw of the Baltimore County PD (cited in Melvin Harris's *Investigating the Unexplained* [2003]) sternly disagreed:

> *Ms. Stokes did not contribute any useful or informative information nor did she supply any new information which could not have been given her by the Griffin family or by newspaper articles printed prior to her visit. Everything she told us was after she had extensive consultations with Griffin relatives and had access to newspaper files collected by the Griffin family.*

THE PSYCHIC MAFIA

M. Lamar Keene had a highly successful career as a phony psychic from the 1950s until his conscience finally caught up with him in the 1970s. He essentially fell in love—platonically, not romantically—with an older, widowed sitter, Mrs. Florence Hutchison, and eventually couldn't stand the pain of deceiving her any longer. When he confessed everything to her, she said she'd stand by him as he told the world of his own and others' fraudulent mediumship. They adopted each other as mother and son, and he changed his name to Charles Hutchison.

After this, he published a book, *The Psychic Mafia* (1976, with Allen Spraggett), in which he not only detailed his own many swindles but also blew the whistle on some of the "psychics" who had been up until then quite highly regarded in the field—such as the late Mable (*sic*) Riffle, who ran the Chesterfield Spiritualist Camp. This was a sort of Disneyland for Spiritualism fans, who could take a vacation staying in a natty hotel while spending their days being swindled by their choice of a passel of fake mediums. The book's publication sparked outrage among Keene's ex-colleagues and brought him death threats and, by his own account, a couple of attempts on his life.

Keene described the absolute contempt that he and his fellow fake mediums had for the suckers from whom they were industriously extracting cash. And it's clear that, even after his damascene conversion, much of that contempt remained. Quite understandably so. There are two contributors to the type of fraud that Keene and his fellows were perpetrating: the fake medium and the sucker. We all joke about the sucker who'll buy the Brooklyn Bridge, but a lot of the stunts Keene and his fellow con artists pulled were at least as blatant as that. The suckers weren't just gulls; they were eager collaborators in their own swindling. Keene characterized the situation as "true-believer syndrome," and spelled out its perils:

> *One of the most alarming things about the mediumistic racket
> is how completely some people put their lives into the hands of
> ill-educated, emotionally unbalanced individuals who claim a
> hot line to heaven. As a medium I was routinely asked about
> business decisions, marital problems, whether to have an
> abortion, how to improve sexual performance, and similar
> intimate and important subjects. That people who ask such
> questions of a medium are risking their mental, moral, and
> monetary health is a shocking but quite accurate description of
> the matter.*

Some of the methods he described as in widespread use are quite embarrassingly crass. How to get hold of information that can later be introduced to amaze the sitters with your "arcane" knowledge? Aside from all the usual techniques, simple pickpocketing is a valuable art to master. You can even filch small objects to be used in later sessions as apports. Best of all is if the sitter at a later session asks the medium for help in locating a small but much-loved item that has inexplicably gone missing. When women put their purses down on the floor during a séance, the medium's sidekick can, in the darkness, sneak the purse to another room and rifle through its contents. Another trick is the use of "stooges." Keene often preferred winos in this role: They were willing to work real cheap and they'd be too eager for repeat paid "bookings" to spill the beans. Even if they did, who would believe them? (You'd have thought, though, that the sitters might complain about those powerful Colt 45 Malt Liquor fumes, and worse, filling the inky black of the séance room.)

A particular trick that Keene described was "precipitation" of "spirit photographs" onto silk. The technique was to take photographs—either of sitters' loved ones or, cut from magazines, of strangers—soak them in ammonia for thirty seconds, and then press them onto bridal silk or satin using a hot iron. Keene recalled that the

⇒ COLD READING ⇐

The most important tool in the toolkit of the phony medium is, without a doubt, "cold reading." This term refers to a set of techniques whereby someone can find out through apparently casual conversation with a stranger far more than the stranger realizes. The precise techniques used vary from one cold reader to the next. Here are just three:

- ◆ Describe common characteristics as if they were specific to the stranger, and be flexible in your response: "When you were a child, there was this puppy you were fond of . . . not a puppy . . . a little dog, then, like a Chihuahua or Pomeranian? Ah, a kitten . . . Oh, yes, I see it clearly now: your baby sister . . ."

- ◆ Offer direct alternatives, and notice the one to which the stranger reacts: "You're tough as nails when you need to be, but really you're a big softy at heart." Most people will make some sign of recognizing themselves in one description or another without registering that you've described two quite different characteristics. (Besides, perhaps they are tough sometimes but also soft at heart, in which case you've struck gold.)

- ◆ Fish: "I'm seeing the letter E. There's someone important in your life associated with the letter E. Could it be a beloved aunt . . .?"

only time he'd had a close shave with this trick was when a sitter recognized the "spirit" on the silk as coming from the cover of a recent issue of *Life*.

The best of all rip-offs, Keene recommended, was "astral development." Here, the medium's spirit guides promised they'd visit the client

while he or she was asleep and provide enlightenment at a time when the person was wide open to the influences of the astral plane, or whatever. Of course, the client might not remember this intervention on awakening, but the benefit would be undiminished. So far as the medium was concerned, this was as near to money for nothing as it got.

❧ THE WILL TO BELIEVE

In his short opera *The Medium* (1946), Gian Carlo Menotti (who wrote the libretto as well as the music) identified a factor that some decades later M. Lamar Keene would describe as "true-believer syndrome" (sometimes referred to elsewhere as "belief persistence"). Madame Flora, a fake medium who's at the heart of the story, finally confesses that she's a fraud. The reaction of her clients is outrage—but not outrage that she's been cheating them. They point-blank refuse to believe her confession.

Daniel Dunglas Home spent some time in his book *Lights and Shadows of Spiritualism* (1877) debunking the claims of his fellow mediums and explaining the methods by which they achieved some of their effects. To produce a spirit manifestation, for example, a medium might wear a modest slip under a dress that could be quickly removed. As soon as she was hidden in the cabinet or behind the curtain, she'd pull off the dress and arrange it so that anyone who caught a glimpse of it would assume they saw the sprawled medium. From her drawers she could produce a handkerchief and tie it around her head, hiding her hair and disguising the shape of her face. Also from her underwear she might pull a thin muslin veil to drape over her body. (The muslin could be folded small enough to be concealed even more intimately than

An 1892 drawing of the medium Eusapia Palladino under investigation. Despite being repeatedly caught in chicanery, Palladino was surprisingly successful.

⇛ FRAUD OR DISSOCIATION? ⇚

On being confronted by the many cases of fraud among psychics, some parapsychologists have pointed out that mainstream science should attend to its own house before criticizing parapsychology. There have indeed been cases of cheating in orthodox science, and it's arguable that not enough is done by the scientific establishment to detect existing cheats and deter future ones. However, as I argued in detail in my earlier book *Corrupted Science* (2007), it's evident that the level of fraud is pretty minuscule in terms of the overall enterprise. There certainly is some fraud in mainstream science, but the structure of the modern scientific enterprise is such that sooner or later—and generally sooner—in any matter of import the fraud is detected and the necessary steps taken to correct the misinformation. (This is what makes claims by politicians that climate change is a hoax perpetrated by climate scientists sound so ludicrous. The conclusions given in any scientist's published work in a field of importance are put to the test by countless other scientists.) It's this integrity of the scientific method that also makes scientists notoriously bad investigators of the paranormal: They're simply not geared to consider the possibility of cheating, because in their ordinary experience it almost never happens. However you evaluate the level of cheating among psychics, on the other hand, it's obvious that fraud is rife.

It's clearly impossible to know how many of the spirit mediums operating in the wake of the Fox sisters were outright frauds. But the best estimates suggest it was most of them, with most of the remainder being cases of genuine self-delusion. In some instances, though, there can be no doubt that the medium went into a trance state and spoke with voices other than the medium's own. While this might have seemed to audience and medium alike to be communication with the Great Beyond, what seems to have been happening was that the medium was going into a dissociative state. In such a state, one's conscious mind is put on the back burner while the rest of one's mental functioning carries on as usual. This is what happens when the mind is in a hypnotic trance, which is not

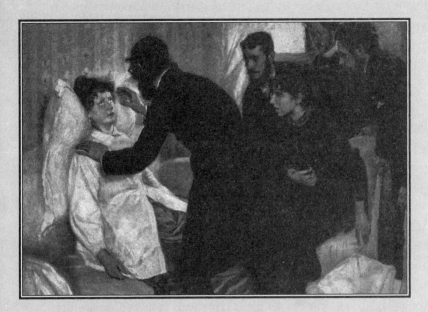

An 1887 depiction of a "hypnotic séance" by the Swedish painter Richard Bergh.

dissimilar to the mediumistic trance state. Dissociation is at the root also of such psychiatric conditions as hysteria and what used to be called "multiple personality disorder" (now drably renamed "dissociative identity disorder"). It is therefore quite to be expected that entranced mediums should speak in other voices, throw their body around, and afterward be unable to recall what went on. A characteristic of less extreme dissociative states is the production of unconscious muscle movements . . . such as those that might rap a table or slide a Ouija board.

Another idea that's often offered is that spirit mediums suffer from dissociative identity disorder and that their controls or guides are in reality their subordinate personalities, using the opportunity of the trance to make themselves heard. This would dovetail nicely with the notion that people who are spoken to by "gods" or other voices in their heads are in fact being spoken to by their own subconscious.

just inside her underwear, if she were worried that the preliminary search might be unusually thorough.) It seems, then, that quite an elaborate series of maneuvers had to be carried out before the "spirit" could emerge from concealment, and there was a similarly elaborate routine afterward. Thus, the fact that sitters didn't more often discern what was really happening suggests that "true-believer syndrome" was as potent in Home's day as it was decades later in Keene's.

In his book *Flim-Flam!* (1980), James Randi recounts how he once recorded the supposed psychic detective Peter Hurkos's appearance on a television talk show and, the following day, listened to acquaintances commenting on how amazing Hurkos's accuracy had been as he psychically revealed various bits of information about audience members. Some days later, Randi asked two of the acquaintances to put on tape their recollections of Hurkos's performance and then played them the recording he'd made of the performance itself. Contrary to their recollections, Hurkos had actually received a hit rate of 1:14. (To stress, that's not 14:1, but 1:14.) Mere random guesswork about the number of children people have or whether they have someone close to them whose name begins with C—and that was about the level of Hurkos's "psychic deduction"—should produce a result at least as good as this. Add in the most rudimentary ability to cold read, and Hurkos, in fact, did very poorly.

So why did people misremember? Why did they credit him with having demonstrated an ability he manifestly did not have? Why did they forget all the mistaken guesses and remember only the occasional lucky strikes? Was it just self-delusion, the satisfying of some inner need to believe in the "spiritual"? Or was it, perhaps, that they were *told* by the program that this was going to be a demonstration of psychic powers, so they unconsciously adjusted their perceptions? Or was it part of the larger psychological tendency to impose order on the world, to narrativize?

In *The Haunted Mind* (1959), the psychoanalyst Nandor Fodor—who was sympathetic to the paranormal—gave a perfect example of how the scattershot statements made by spirit mediums can be given significance by the sitters. In 1933 he attended a séance held in London by the medium Mrs. Heath. She threw a few names at him that were supposed to be of people important to him; none were. She made some predictions about him—including that he too would become, in due course, a spirit medium—none of which had come true by the time of his writing, a quarter-century later. And then, perhaps assuming that Fodor,

Czar Nicholas II of Russia: Did he really convey personal thanks from the Great Beyond to a Hungarian psychoanalyst?

a Hungarian, was a Russian, she talked about snowy wastelands and informed him that none other than Czar Nicholas II of Russia was here in spirit to convey his personal thanks for something Fodor had done to aid the Romanov family.

At the time, this meant nothing to Fodor—just another wry tale to tell at the dinner table. But, not long after, he read of the death of Grand Duke Alexander of Russia, Nicholas's brother-in-law, and remembered something he'd long forgotten. One day, while Fodor had been working for the British newspaper publisher Lord Rothermere, Alexander had sent his private secretary with a request for an appointment to see Rothermere, and Fodor had agreed to pass the request along.

As Fodor describes it, for a few glorious moments he thought the consequences of that small action might have been considerable for the surviving Romanovs, sufficiently so for Nicholas to want to make a personal gesture of gratitude . . . But then his rational mind kicked in, and he realized he was letting his desire to believe lead him into making a false correlation. After all, working as he did on Rothermere's secretarial staff, it was his job to meet people like the private secretary to Grand Duke Alexander; the connection represented no astonishing coincidence. Yet his mind had been eager to construct a plausible narrative out of the two unrelated events.

2

HAUNTINGS, HALLUCINATIONS, OR PSYCHOKINESIS?

PEDDLING POLTERGEISTS, EVP, SOUL-STUFF, AND MORE

"The accumulation of specters of the different tribes of the terrestrial fauna, heaped at the surface of the globe since the first geological epochs, would render the air irrespirable. We could not move, in a dense atmosphere of ghosts."

—Adolphe d'Assier, *Posthumous Humanity* (1887)

O NE FACTOR THAT MAKES IT DIFFICULT TO EVALUATE the classic accounts of ghosts and hauntings that appear in the literature is that they differ between one telling and the next, and sometimes in fundamentally impor-tant ways. To take just a single example, there's the tale of the Chaffin family of Mocksville, North Carolina, during the 1920s.

As recounted by Colin Wilson in *Afterlife* (1985), James Chaffin Sr. died in the early 1920s, leaving everything to the third of his four sons, Marshall. In June 1925, the old man's ghost appeared to one of the other sons, James Chaffin Jr., telling him that a revised will could be found in the lining of an inside pocket of an old black coat that he, the father, had habitually worn. James Jr. traced the coat to the house of his brother John but, when the two opened up the lining, they found not a will but the instruction to go check Genesis 27 in the family Bible.[1] There, between two pages of the neglected book, they found a will, dated later than the one that had given everything to Marshall, decreeing that the estate be divided equally among the four sons and their mother. "The first reaction of Marshall Chaffin was to contest the will," writes Wilson, "assuming it to be a forgery. But, once he examined it, he had to admit that it was obviously genuine."

---

[1] In Genesis 27, Jacob swindles his brother Esau out of his inheritance by deceiving their blind father, Isaac.

The same case is discussed by Mary Roach in her book *Spook* (2005). For the most part, her account matches Wilson's, but the details in which it differs are significant. Marshall, we learn, died within a year of his father, so that, by the time the ghost appeared, Marshall was out of the picture. He couldn't have agreed that the new will was "obviously genuine" for the very good reason that, by then, he'd been dead for three years. Marshall's widow, Susie, by contrast, was still very much alive and she was determined to contest the newly discovered will, which she believed to be a forgery—as it almost certainly was, according to a handwriting expert Roach hired to examine the document. Even so, the brothers were able to muster a number of witnesses who were prepared to testify that the signature on the will was, indeed, James Chaffin Sr.'s. Presumably, many of the locals felt that a great injustice had been done and perhaps that Susie Chaffin—who was by all accounts a formidable woman—had bamboozled the old man into leaving his entire estate to Marshall.

On the opening day of the trial, during the lunch break, the surviving brothers hammered out a deal with Susie whereby the estate was divided equally between them, and the case went no further. It seems very likely that Susie, realizing that, whatever the validity of this new will, she was quite likely to lose the court case, sensibly decided to settle for what she could get.

In light of the information that Roach gives, the "ghost story" part of the episode suddenly seems exceptionally artificial, as if it might have been concocted purely in order to offer some "plausible" reason for the delayed discovery of a forged will. Whatever the truth of the matter, what seemed initially like a valid instance of a ghostly manifestation is now clearly anything but. We can't rely upon it as evidence either way regarding the existence of ghosts.

❧ BORLEY RECTORY

Reverend Henry Bull built Borley Rectory in the county of Essex, in the English southeast, in 1862. There were sporadic tales of vaguely spooky happenings over the next few decades, both while Bull was alive and when his son—another Reverend Henry Bull—took over as vicar. In 1928 the younger Henry died, and some time later Reverend Guy Smith and his wife moved in. Then the haunting began in earnest. The Smiths contacted the newspaper the *Daily Mirror*, hoping to be put in touch with the SPR.

The Reverend Henry Bull Sr., who built Borley Rectory, dying there in 1892.

It's at this point in the story that the first alarm bell rings. Why on earth would the Smiths think the best way of getting in touch with the SPR would be to contact a downmarket newspaper? A London telephone directory—surely standard issue at the local library—would have been a better bet. It's at least possible that, remuneration for Church of England incumbencies being notoriously skimpy, the Smiths were hoping to make a little money out of their "haunted" house.

The *Mirror* sent along a reporter, and, in due course, the psychic researcher Harry Price arrived. While Price was there, the "phenomena" picked up in intensity, but when he left they reverted to their normal level, so the Smiths suspected he had been pulling a fast one. In her history of the SPR, Renée Haynes remarks:

> Price was uninhibited by a sense of honour, general or scholarly;
> and he combined an infinite capacity for drama with a great

talent for, and addiction to, publicity, sometimes of a ridiculous kind, as when he associated himself with the story of a talking mongoose called Jeff, said to appear from time to time on the Isle of Man and to speak—and sing hymns—in six European languages (thus distinguishing himself from the animal "familiars" of folklore witches, who had an aversion to all things ecclesiastical).

The Smiths didn't last long at Borley. The next residents were Reverend Lionel Foyster and his much younger wife, Marianne. At this stage—if we're to believe the stories—the ghost or ghosts of Borley Rectory really began to let rip: Windows shattered, bells rang, objects were thrown around and, at least once, Marianne was decanted from her bed by a mysterious force. A couple of exorcism attempts, carried out by Foyster himself, failed. Around 1935, Foyster wrote a rambling account of all this and sent it to Price, who again investigated and then declared the rectory was . . . well, the title of his first best seller on the subject says it all: *The Most Haunted House in England* (1940).[2] The Foysters departed in 1935, and the house was largely destroyed by a fire in 1939.

The first substantial debunking of Price's work came in 1956. In *The Haunting of Borley Rectory*, Eric Dingwall, Kathleen Goldney, and Trevor Hall revealed that many

The elaborate kit that Harry Price took with him on his various ghost-hunting expeditions.

---

[2] He later added a further book on the subject, *The End of Borley Rectory* (1946). Another account for the credulous was *The Ghosts of Borley* (1975) by Paul Tabori and Peter Underwood.

Borley Rectory is no more. The place suffered a fire in 1939 and the remains were eventually demolished in 1944.

of the "supernatural" phenomena had been produced by Price himself (he was, after all, a skilled amateur conjurer). Many others were devised by Marianne Foyster. When the book was reviewed in the *Economist*, the reviewer made remarks that are still pertinent today:

> *Entertaining as the book is, it leaves a slightly nasty taste in the mouth. Not because the late Harry Price emerges unmistakably as a rogue, a falsifier and manufacturer of evidence; not because certain other persons, with motives less material though understandable enough, faked a bigger volume of evidence than he did; but because the whole long drawn-out Borley affair at once constituted a debasement of popular opinion and thought. Everything that was shoddy, muzzy, slipshod and anti-rational in the public mind responded to, and throve on, the Borley sensation. It will take*

more than this antidote to counter so massive and thoroughly
assimilated a dose of dope.

In other words, irrational beliefs might seem harmless—even humorous—but they contribute to a public brainrot that can have very dire consequences.

Central to the story of Borley Rectory is the image of the rectory's tenants during the years from 1930 to 1935, the time of the greatest spectral activity: the staid, respectable Reverend Foyster and his devoted younger wife, Marianne. By dint of some extraordinarily patient detective work, Trevor Hall was able to piece together a history of this curious couple that differed quite a lot from the official one.

When Marianne was fifteen, she got pregnant by and married a man named Harold Greenwood in her native Ireland. Soon afterward, Greenwood fled, leaving Marianne and her parents to rear the child, Ian. In 1922, when Marianne emigrated to New Brunswick, Canada, she took her son with her, pretending he was her kid brother. In Canada, she (bigamously) married Reverend Foyster, and some years later, when Foyster lost his incumbency in New Brunswick, the family moved to England to take up the living and the rectory at Borley. They took a tenant for the rectory's cottage. This man, Frank Pearless (aka Françoix d'Arles), became Marianne's lover. It seems, too, that as well as rather openly cuckolding the elderly Reverend Foyster, Pearless enjoyed playing sadistic little tricks on him, often with Marianne's connivance. Foyster was all too gullible: According to an account Ian Greenwood gave to Trevor Hall, eventually Pearless felt things had gone too far and told Marianne they must stop the pranking.

But the reverend wasn't entirely naive; he was sufficiently aware of the largely imaginary nature of the ghosts of Borley that he conceived the lucrative idea of writing a fraudulent account of life at the rectory, full of as many spooky accounts as he felt readers might

➤ THE AMITYVILLE HORROR ⬅

Another very famous haunting—that of a house in Amityville, New York—is one of the most thoroughly debunked ghostly hoaxes. The true part is that, in late 1975, George and Kathy Lutz, with Kathy's children, moved into a house in Amityville where, just over a year earlier, Ronald DeFeo had murdered his parents, his two brothers, and his two sisters. DeFeo's defense, likely genuine (he suffered from some kind of schizophrenic disorder), was that he "heard voices plotting against him," but he was nevertheless convicted on six counts of second-degree murder rather than committed to psychiatric care.

What supposedly happened thereafter is described in Jay Anson's *The Amityville Horror: A True Story* (1977), which was made into a sensational movie that spawned eleven sequels. The Lutzes lasted in the house for about a month before being driven out by all the spooky activity—the demonic cacklings, the oozing slime, the . . . well, fill in your own cliché here. Around the time that the movie was released, the Lutzes sued DeFeo's defense lawyer, William Weber, and others over various media articles that had appeared about the case. This proved to be a mistake, because not only was their suit thrown out, but Judge Jack B. Weinstein declared,

swallow. He knew there could be money in such a book because, while living in St. John, New Brunswick, he had read the best seller *The Haunted House* (1879) by Walter Hubbell. This book, although published in remote New Brunswick about a haunting in even remoter Nova Scotia—the "Amherst Mystery"—managed to sell over fifty thousand copies.

After the Foysters left Borley, yet another moneymaking scheme occurred to them. In 1935, Lionel Foyster pretended to be Marianne's father when she lured a commercial traveler named

"Based on what I have heard, it appears to me that to a large extent the book is a work of fiction, relying in a large part upon the suggestions of Mr. Weber." Weber confirmed this in an article for *People*: "I know this book is a hoax. We [Weber and the Lutzes] created this horror story over many bottles of wine." Further lawsuits followed. The

next occupants, Jim and Barbara Cromarty, found many discrepancies in the Lutz/Anson description of the house and maintained that the only thing haunted about it was that rubberneckers kept turning up on their doorstep. Their invasion-of-privacy suit was successful, as was that of Father Ralph J. Pecoraro (called "Mancuso" in Anson's book and "Delaney" in the movie), who supposedly tried to exorcise the house and was tormented by demonic attempts on his life as a result. Pecoraro insisted that the nearest he came to the house was a phone call with the Lutzes (although he later appeared to backtrack on this a bit).

Henry Francis Fisher into yet another bigamous—trigamous?—marriage. Until the reverend's death in 1945, they lived with Fisher, maintaining the pretense. By then, Marianne had a new lover: an American GI stationed in England. Around the time of Foyster's death, Marianne persuaded the lover that she'd gotten pregnant by him and, while Fisher was absent on a business trip, entered into yet another illicit marriage. After the GI departed for home, she was able to adopt an infant and followed him to the United States as his bride.

✣ PLOTTING POLTERGEISTS

In modern studies of the psychic and parapsychology, poltergeists are most often regarded as being a form of involuntary psychokinesis (PK)—the purported power of the human mind to move objects around or effect other physical changes in the surroundings through sheer willpower alone. Uri Geller's celebrated cutlery-bending is a prime example of this hypothetical ability. Until the latter part of the twentieth century, however, poltergeists were most commonly assumed to be ghosts; the term *poltergeist* itself comes from the German for "noisy ghost" or "banging ghost."

The granddaddy of all poltergeist cases is arguably the Cock Lane Ghost, which was the talk of London in 1761–1762. Following the death during childbirth of his wife Elizabeth, William Kent, a Norfolk man, took Elizabeth's younger sister, Fanny Lynes, as his common-law wife. The pair moved to London, where they rented rooms in Cock Lane, near St. Paul's Cathedral, from the parish clerk Richard Parsons. After complaining of odd noises and other phenomena, the couple moved out, and the activity stopped. But, after Fanny died of smallpox, the rappings resumed, seemingly focused in the wall behind the bed of Parsons's ten-year-old daughter, also named Elizabeth. Parsons called in a local preacher, the Reverend John Moore, who managed to open communication with the "ghost" using a one-knock-for-yes, two-knocks-for-no code. Through this means, Moore and Parsons were able to uncover a horrifying secret: Fanny Lynes had died not of smallpox but from poisoning by arsenic, and her murderer was William Kent!

This was a great tale, and Cock Lane drew curiosity-seekers from all over London in hopes—often fulfilled—of witnessing the poltergeist's manifestations. But one relevant part of the story had been oddly underemphasized. On arrival in London, Kent had lent Parsons twenty pounds, and, in due course, was forced to file a lawsuit in order to get the loan repaid. Parsons seems to have nurtured a powerful

Detail from satirist William Hogarth's 1762 print *Credulity, Superstition and Fanaticism,* **which railed against the climate of irrationality. The Cock Lane ghost is shown in the form of a carving being slipped by a preacher into a girl's dress (right). The sprawled woman represents Mary Toft, who in a notorious 1726 cause célèbre claimed to give birth to multiple rabbits.**

➤ THE ENFIELD POLTERGEIST ⇐

The Enfield Poltergeist case, the subject of Guy Lyon Playfair's book *This House Is Haunted* (1980), occurred in 1977–1979 in Enfield, a suburban borough of North London. The affected family consisted of separated wife Peggy H. and her four kids: Margaret, twelve; Janet, eleven; Johnny, ten; and Billy, seven.[1] It displayed the traits typical of a poltergeist manifestation: furniture being moved around, rattlings and rumblings, and—most dramatically of all—the wrenching of a heavy fireplace from the wall. In addition, there was a ghostly, foul-mouthed voice that spoke through Janet's false vocal cords (ventricular folds). This voice claimed to be that of one Bill Wilkins, who had died in the house many years before. The overall effect was not unlike the demonic vocalizations of Linda Blair's character in the movie *The Exorcist* (1973).

The case was investigated by the SPR's Maurice Grosse and Guy Lyon Playfair, and it made national headlines. The two investigators, with photographer Graham Morris in tow, were quite convinced there was a genuine supernatural phenomenon at work here—even though, from time to time, they caught the two girls, Margaret and Janet, cheating (the pair even went so far as to admit to TV reporters that they cheated "some of the time"). Morris captured some famous photographs of flying objects—including, on occasion, Janet appearing to levitate.

In July 1978, Janet was admitted to the Maudsley Hospital in South London for psychiatric evaluation. By the time she returned home a couple of months later, having been given the all clear, the "manifestation" was on its last legs. She claimed she had experienced some minor poltergeist activity while in the hospital, but there's no independent confirmation of this. Playfair concluded his account with this observation:

When Mr. and Mrs. [H.] were divorced, an atmosphere of tension built up among the children and their mother, just at the time when the two girls were approaching physical maturity. They were a very energetic pair to start with, both of them school sports champions, but even they could not use up the tremendous energy they were generating. So a number of entities came in and helped themselves to it.

Looking skeptically at the situation, parts of Playfair's analysis make sense. The two girls probably did have more energy than they could use up through normal channels . . . and it seems like a pretty obvious conclusion what they used that energy for. The famous photos of Janet being thrown across the room by the poltergeist look for all the world like photos of an athletic little girl throwing herself out of bed. And, as with so many modern poltergeist cases, the investigators noticed that the physical manifestations never seemed to happen when they were looking in that direction. For instance, after Playfair asked Janet if she could bend cutlery, like Uri Geller, he turned away for a moment to speak to Peggy. When he turned back—presto!—there was a bent spoon on the kitchen table.

Decades later, in 2015, the case re-entered the public consciousness thanks to the British miniseries *The Enfield Haunting*, starring Timothy Spall as Maurice Grosse and Matthew Macfadyen as Guy Lyon Playfair. It's an extremely entertaining and neatly made piece of work, but it has about as much documentary veracity as Geraldo Rivera's *Al Capone's Vault* broadcast.

---

[1] In his book on the case, Playfair changes some of the names of those involved.

resentment over the whole affair. Eventually, it was realized that the desire for revenge might be playing a part in the proceedings and, when Elizabeth was caught cheating, the assumption was that her father had put her up to the whole performance. Parsons was sent to the pillory three times and sentenced to two years in prison—a remarkably lenient sentence, considering that he'd been trying to get Kent hanged.

EARTH TREMORS OR NERVOUS ENERGY?

The SPR's Frank Podmore investigated a dozen poltergeist cases around the end of the nineteenth century and concluded that eleven of them were based purely on trickery; he thought it possible that some form of the supernatural was involved in the twelfth.

Various theories to explain poltergeist phenomena emerged within and outside the SPR. As we saw on page 48, in the nineteenth century the geologist John Milne initiated the notion that a lot of these events might be explained by earth tremors. In the 1950s, this was picked up by the SPR and expanded: Perhaps earth tremors might also frighten people into hallucinations of ghosts. And what about other physical explanations? Might luminous specters be patches of phosphorescent gas arising from damaged sewers? (This theory explained the fact that sightings of these apparitions were much more common before the old sewers had been repaired and rebuilt.)

A study by Guy W. Lambert, published in the SPR's *Journal* in 1960, demonstrated that many of London's supposedly haunted houses stood near subterranean watercourses, like the Tyburn Brook. Running water dislodges rocks and chunks of soil or rolls objects along the streambed, and many mysterious bumps in the night might perhaps be traced to this.

The geophysical explanation was just a passing fad. In 1961, the parapsychologists Alan Gauld and A. D. Cornell were able to borrow a house scheduled for demolition; they placed objects on shelves and window ledges within, and then applied heavy vibrating equipment to the outer walls. By the time cracks were starting to appear in the walls, one or two of the smaller items had been shaken off their shelves, but there was none of the hurtling around of objects typically reported in poltergeist cases.[3]

As we've noted, much more important to the story of parapsychology has been the notion of poltergeist activity as a form of involuntary psychokinesis (PK); indeed, parapsychologist J. Gaither Pratt suggested that the poltergeist phenomenon should be renamed RSPK for "Recurrent Spontaneous Psychokinesis." Usually the PK is traced to a pubescent child, and the assumption is that the wild energies of puberty are somehow spilling out of the body to manifest themselves as unearthly chuckles and the smashing of family possessions—rather like ordinary adolescence, in fact. The German parapsychologist Hans Bender had a more physics-based notion of where those wild energies might come from, pointing out that a drop in room temperature of just a single degree represents a really quite large amount of energy—so what the poltergeist agent was actually doing was draining energy out of the air and, in the process, cooling it. This cooling would not only offer a ready local source of energy for the poltergeist activity but also explain the chill that accompanies ghostly presences in many traditional accounts.

In an April 1948 essay in *Psychiatric Quarterly*, "The Poltergeist—Psychoanalyzed," Hungarian-born parapsychologist Nandor Fodor posited that the supposed psychological disorder at the root of

---

[3] Also of interest is that they reported no uncanny sensations as a result of all the subsonics that must have been filling the house's interior in addition to the ordinary racket. For long it was thought that subsonics—sounds too low to hear—might be responsible for many instances of "creepy feelings" later reported as supernatural. It's not known why Gauld and Cornell didn't experience these sensations, since accounts of the correlation are quite widespread.

pubescent PK was, as he put it much later in his book *The Haunted Mind* (1959), "an episodic mental disturbance of schizophrenic character." Further ideas of Fodor's were summarized by the American paranormal researcher D. Scott Rogo in *On the Track of the Poltergeist* (1986): "The poltergeist may be masterminded by some portion of the agent's unconscious mind that has physically detached itself from the host's brain, psyche and body and has developed a primitive intelligence of its own."

PK FAKERY: A CRY FOR HELP

In 1984, not long after their fourteen-year-old adopted daughter Tina had watched and been much affected by the movie *Poltergeist* (1982), the Columbus, Ohio, home of John and Joan Resch was assailed by all sorts of poltergeist manifestations: flying objects, lights flashing off and on—the usual. Journalists were permitted to come into the home and marvel at these uncanny doings, as was the parapsychologist William Roll. Oddly enough, though, when a three-strong investigatory team from CSICOP[4]—astronomers Steve Shore and Nick Sanduleak, plus conjurer James Randi—showed up, the visitors were refused entry.

A TV crew did better, although the poltergeist seemed oddly shy whenever the camera was focused on Tina. However, when the team left the room at one point, the activities suddenly started up at full throttle. Alas for Tina, the crew had inadvertently left the camera on, and later, examining their footage, they were able to see exactly how Tina had created her effects. Confronted by the evidence, which was broadcast on TV, Tina's mom insisted that Tina had been tired and merely wanted everyone to go away,

---

[4] The Committee for the Scientific Investigation of Claims of the Paranormal (CSICOP) was founded on April 30–May 1, 1976. It was later renamed the Committee for Skeptical Inquiry (CSI).

cheating just this once. But if the Resches were so convinced they'd genuinely been invaded by a poltergeist, why did they bar the CSICOP team? Shouldn't they have been eager to gain the support of these hardened skeptics?

Much of the coverage of the case was done by the *Columbus Dispatch*, whose reporter, Mike Harden, had written about the Resches on other occasions: John and Joan Resch had fostered more than 250 children over the years—an impressive contribution to the community. The newspaper's photographer Fred Shannon took the pictures that were soon appearing on front pages all over the nation. Randi, denied direct access to the Resches, was allowed—after some argument—to purchase from the *Dispatch* a contact sheet of Shannon's photographs, including those that the paper had declined to use. Later, when Randi prepared his report on the case for *Skeptical Inquirer*, the newspaper refused permission for the magazine to print the photos alongside Randi's article; it had to make do with artist's sketches closely based on the photos. The reason for the newspaper's caginess was, of course, that in some instances the photos could more readily be interpreted to show Tina cheating than as evidence of poltergeist activity—and where's the story in that?

Shannon still maintains that the photos show supernatural phenomena—and, after all, he was there. He does, however, admit to a curious difficulty when taking the photographs: The phenomena in question never happened when he was actually looking at them—he always saw them as a sort of blur at the edge of his vision. Thus he learned that the best method was to point the camera at Tina, look away, and press the shutter button when he saw that blur. While to us this might suggest deception, to Shannon it indicated that whatever occult power was involved was leery of direct observation. Perhaps.

Years later, in 1994, Tina—who had by now changed her name—was convicted of the murder of her three-year-old daughter, Amber. In fact she entered a so-called Alford plea, whereby the accused is allowed to plead guilty in order to avoid harsher punishment while yet maintaining his or her innocence. Of course, this plea is distinctly Lewis Carrollesque (welcome to the US judicial system!). She is still in prison.

In all the literature on poltergeists, there's a recurring theme: The activity focuses on a child (or sometimes children) prior to or during the onset of puberty. Usually, but not always, the child is a girl. The same general picture holds true among certain Spiritualist phenomena—such as the Hydesville rappings (page 2)—and with many purported cases of demonic possession, including the one that inspired William Peter Blatty's novel *The Exorcist* (1971) and its 1973 movie adaptation. Children who are unhappy for one reason or another—perhaps they feel neglected or the family has been disrupted—often fall into a pattern of attention-getting behavior. Also, children are far smarter and more cunning than adults often realize—just think of all those thirteen-year-olds whose passion in life is performing magic tricks. PK, after all, is just another form of conjuring.

One of the most out-of-left-field explanations for poltergeist activity came from British physicist John Hasted in *The Metal-Benders* (1981), which offered an account of personal and allegedly scientific investigations into psychokinesis—particularly the cutlery-bending made popular by Uri Geller. He noted that poltergeists in Victorian-style houses had a propensity for ringing doorbells and those servants' bells you see in *Downton Abbey*—the type controlled by metal wires. Hasted suggested that the metal wires might offer a clue. Could someone in the house have been unconsciously bending those wires in Uri Geller fashion?

The mathematical physicist John G. Taylor was more than a little skeptical of Hasted's theory, offering the following wisdom at the end of his (unfavorable) review of Hasted's "monumentally silly" book in *New Scientist*: "Believe nothing that you hear and only half of what you see."

❧ SPOOKY PSYCHOLOGY

In the 1930s the Oxford philosopher H. H. Price (not to be confused with the Harry Price of Borley Rectory fame) offered up the notion that there might be some kind of psychic ether—slightly analogous to the physicists' luminiferous ether and, like it, permeating the entire universe. A property of the psychic ether, Price speculated, would be that it was capable of being imprinted by mental images—the stronger the mental image, as in a moment of terror, the deeper and more permanent the imprinting. Since most supposed hauntings involve people who've met a violent end, this would seem logical enough. The idea that ghosts aren't really "restless spirits" but merely "tape recordings" of past events has gained considerable popularity ever since Price proposed it, although, of course, there's absolutely zero evidence for the existence of a psychic ether.

More firmly grounded in reality, some experiments that might explain a lot of ghost sightings (and a lot of UFO close encounters, for that matter) were carried out in the 1970s by the psychiatrist Morton Schatzman. A patient whom he called Ruth came to him because she was tormented by apparitions of the father who had cruelly abused her in childhood. She'd have thought these apparitions were his ghost except that he was still alive—although, mercifully, far away. Nevertheless, the apparitions were very real to her. If the specter touched her, she felt his touch, and sometimes she could smell the booze on his breath.

That the apparitions were born of Ruth's mind, rather than of supernatural origin, was apparent: As just a single example, since last she'd seen him, Ruth's father had grown a beard, but the "apparition father" was beardless. Intrigued, Schatzman worked with Ruth until she learned how to summon apparitions at will—and not just of her father but of anyone else she chose. Once they appeared, the apparitions seemed to have autonomy. She could hold meaningful conversations with them as they moved around of their own volition. In one memorable instance, she had sex to the point of orgasm with the apparition of her husband. Most interestingly of all, when Schatzman hooked her up to an electroencephalograph and sat her in front of an image that induced a particular pattern of brain waves, the pattern was disrupted every time one of her apparitions "passed in front of the screen"—exactly as if Ruth's view of the screen had indeed been blocked by something.

Before there were psychiatrists—and encephalographs—Ruth's apparitions would have been taken as genuine visitations from the spirits or, if those visitations were from people who were still alive, as "phantasms of the living," to use the SPR's terminology. We have no way of telling how many ghost encounters of the past were born from this psychological phenomenon. Since we also don't know how common the phenomenon is, we can't tell how many modern paranormal experiences are attributable to it, either.

HYPNAGOGIC DREAMS, MISPERCEPTIONS, AND EVERYDAY HALLUCINATIONS

Typically, ghostly events happen when you're drowsy—in bed and just falling asleep, perhaps—and this leads to a likely explanation for many spectral visions. As we're falling asleep, we can experience what are called "hypnagogic images": extremely realistic images that blend in with our natural surroundings and that the mind, in its suggestible state, readily accepts as genuine. Some people undergo hypnagogic dreams habitually—my mother was one—whereas for others it's only an occasional event. (The visions can also occur as we slowly awaken, in which case they're described as "hypnopompic.") The images can be accompanied by sound effects, appropriate or inappropriate. My mother vividly recalled being visited by a cute little lamb that stood on the bed and lectured her at length. But more often the sounds tend to be dissociated whispers, distant bangings, and the like—all the usual paraphernalia of a haunting.

The stage of sleep in which hypnagogic dreams occur is known as stage 1. As you fall deeper into slumber, you pass through stages 2 and 3 until reaching stage 4, the deepest sleep of all, when your brain is at its most idle. But stage 4 sleep doesn't last long. After perhaps a half-hour, it stops and your brain suddenly tumbles through stages 1 to 3 again before entering the condition known as REM sleep—so named because it's characterized by rapid eye movements. It's also characterized by dreaming, sexual arousal, and bodily paralysis: The brain (more accurately, the brain stem) shuts down most muscular movements in case you thrash around and injure yourself in response to your dream.

Sometimes the standard pattern doesn't work just right. We've all seen people twist and moan in the grip of a bad dream, and on other occasions we can—frighteningly!—fully awaken while still locked in the paralysis of the dreaming state. More importantly, in the context of ghostly encounters, it's possible to find ourselves in both the dream state and stage 1 sleep at the same time: We're pseudoconscious and receptive to frighteningly realistic images, we're sexually aroused, and we're paralyzed, as if something suffocatingly heavy were sitting on our chest.

Sound familiar? Even if you haven't experienced this situation yourself, you may recognize it from the famous painting *The Nightmare* (1781) by the Swiss artist Henry Fuseli. On the chest of a restlessly sleeping woman squats a compact and obviously heavy demon. What Fuseli's painting portrays is exactly the situation just described—of being caught in a tangle of stage 1 and dream-state sleep. It's clear that the demon is intended to be an incubus—a malevolent male spirit—usually fearsomely endowed, whose practice is to have intercourse with sleeping women. (Male sleepers are likely to encounter female equivalents called "succubi.")

Of course, those spooky lights and sounds may have far more mundane sources: the headlights of distant cars, ordinary bedroom

This c. 1791 rendition of Henry Fuseli's painting *The Nightmare* offers a graphic representation of nighttime paralysis.

objects misperceived (I can't count how often I've woken up to find myself being observed by a dark and silent figure, only to realize, once my heart has stopped pounding, that it's just my dressing gown), stray reflections in windows or mirrors, differential heating and cooling of the house. All these effects and many more can startle even the most pragmatic of us. If our default assumption is that there must be a natural explanation, then the experience is no more than that: momentary. For others, though, that sighing water pipe in the walls is, yes . . . a ghost!

Finally, hallucinations are a far more common phenomenon than we tend to believe, and they're often extremely hard to distinguish from reality. Obviously they can be brought on by exhaustion, mental or physical illness, drug abuse (or even just ordinary use of prescription drugs), hunger, and so on, but in some instances they just happen. They can also be "infectious": Because we're all so suggestible, if one person sees a ghost in a shadowy corner, chances are that at least some of us will likewise see something *lurking* there.

PHILIP AYLESFORD, SKIPPY, AND THE PHANTOM VICAR OF RATCLIFFE WHARF

The 1976 book *Conjuring Up Philip* by Iris M. Owen and Margaret Sparrow recounts one of the most fascinating experiments in the history of parapsychology. In 1972, the former math professor George Owen, director of the Toronto Society for Psychical Research, convoked a panel of eight volunteers, headed by George's wife, Iris; Margaret Sparrow was one of the other participants. The researchers invented a historical character named "Philip Aylesford" and gave him a biography: He was born in 1624 in England, fought in the Civil War, spied for the Royalists, fell in love with a gypsy girl who was later accused of witchcraft and burned at the stake, and so on. They also gave him a specific appearance. According to George Owen, "It was essential to [our] purpose that Philip be a totally fictitious character. Not merely a figment of the imagination but clearly and obviously so, with a biography full of historical errors."

The panel then used the methods of the séance room in an attempt to get in touch with this spirit—to *create* an apparition or ghost, in fact. And in due course they (apparently) succeeded. As the eight sat around a table, hands upon it, they'd ask "Philip" yes/no questions, and the "spirit" would respond with one rap for yes or two for no; if someone accidentally asked a non-yes/no question, "Philip" would make an irritated scratching noise until they got it right. The "ghost" had a bit of a temper, too. On occasion he'd tilt or twist the table to show his annoyance. This all sounds very much like the ideo-motor effect that can be responsible for table rappings and tiltings in conventional séances or for events that are more generally ascribed to PK, although we're assured that the experimenters took the precaution of putting paper doilies under the participants' hands.

The experimenters' conclusion was that they'd succeeded in demonstrating that Spiritualist phenomena were real but born not from the afterworld but from the human mind. This explanation meshed with other parapsychological thinking—notably that poltergeists are not ghosts but psychological phenomena.

A couple of movies were (very) loosely based on the Philip experiment: *The Apparition* (2012) and *The Quiet Ones* (2014).

There have been attempts to replicate the Philip experiment—notably the Skippy experiment, conducted in Sydney, Australia, by a team under the leadership of paranormal researcher Michael Williams. In this instance, the invented character was fourteen-year-old "Skippy Cartman," who was strangled by her Catholic schoolteacher, with whom she was having an affair, after he discovered she was pregnant. Alas, although great claims have been made for the experiment's success, apparently no video or audio of the results has ever been released—something the Philip experimenters had been able to do—and by about 2007 the reports petered out.

In a similar vein—although perhaps more lightheartedly—in 1970 Frank Smyth and his friend John Philby had the idea of inventing a ghost, spreading rumors about it, and seeing what happened. At the time, they were sitting in a pub in London's Docklands area, near Ratcliffe Wharf. Enter the "Phantom Vicar of Ratcliffe Wharf." According to the newly minted legend, this venomous cleric hired prostitutes to lure sailors to his house, where he'd rob and murder the men before throwing their bodies into the Thames River. Smyth then published the story in the partwork publication *Man, Myth & Magic*, where he was a staff writer.

It didn't take long before the locals were buzzing about the spectral vicar, and quite a few of them reported having graphic encounters with him. In 1973, when Colin Wilson was making his TV series *A Leap in the Dark*, one episode focused on the ghost, and included plenty of eyewitness testimony from the residents. (Since the Phantom Vicar was potentially good for local businesses, it's conceivable that some of these accounts were themselves invented, rather than delusional or the product of suggestion.) Even after Smyth confessed the hoax, in a 1975 interview with a national newspaper, the *Sunday Times*, it was still possible to find locals who disbelieved the confession, "remembering" quite clearly how they'd first been told of the legend in their childhood.

❧ ELECTRONIC VOICE PHENOMENA

Electronic Voice Phenomena (EVP) are sounds picked up via electronic media—typically on blank tapes or dead phone lines—that auditors interpret as voices, usually of the dead, sometimes of aliens, sometimes even of angels or demons. A few theorists have posited that the voices are, rather, examples of unconscious PK on the part of, perhaps, the experimenters themselves. Most skeptics attribute the supposed sounds to natural phenomena, like white noise (page 111).

Sometimes the term Instrumental Transcommunication (ITC) is preferred.

Founded by Sarah Estep in 1982, the American Association of Electronic Voice Phenomena, later renamed Association TransCommunication, has a mission to help people learn about and experiment with EVP. In May 2000, she passed the leadership of the association to Tom and Lisa Butler. Estep had first become interested in EVP in the 1970s, when she discovered that she could hear voices while playing tapes on her husband's tape recorder. After Estep died in 2008, Sonia Rinaldi (page 111) contacted her spirit by telephone. Estep was able to come through with such important messages as "I hear, Sonia" and "Now goodbye."

Interest in the topic goes back much further than the 1970s, though—arguably to the 1920s, when Thomas Alva Edison said in a *Scientific American* interview that, if indeed something of us survived death and hung around trying to communicate with the living, the way to detect their communications was not through table-tilting or spirit mediums but through a sensitive sound-detection apparatus. His purpose in making the remark seems to have been more to brag about his own inventive genius than to suggest that he had any serious intention to build such a device.

In 1957 the painter and opera singer Friedrich Jürgenson bought himself a tape recorder to help his singing practice. As he played back his tapes, he began to notice some odd phenomena—not just auditory but also visual, even telepathic. In the spring of 1959, one of these communications told him about "a Central Investigation Station in Space, from where they conducted profound observations of Mankind":

> *My friends spoke about certain electro-magnetic screens or*
> *radars, that were frequently transmitted, day and night, in*
> *thousands to our three dimensional life levels and like living*

beings had a mission as mental messengers. Undoubtedly one could see these radars as half-living robots that, remote controlled, had the ability like an oversensitive television or radio to correctly register and transmit all our conscious and unconscious impulses, feelings and thoughts.

A few weeks later he recorded some wild birdsong with the machine, and again there was more on the tape than there should have been. This time a voice spoke to him in Norwegian (Jürgenson was of Scandinavian parentage), although what it told him was of little consequence. Initially, he thought these and other voices were those of his extraterrestrial friends, but he began to suspect they were, rather, the voices of the dead—a suspicion confirmed when one day he heard the voice of his deceased mother. He compiled these observations and the results of other researches into a book *Rösterna från Rymden* (1964; translated as *The Voices from Space*).

One person to be strongly influenced by Jürgenson's work was the Swedish writer Konstantin Raudive, who investigated EVP for the last decade or so of his life, making over one hundred thousand audiotapes. It's for him that EVP is sometimes referred to as Raudive Voices, although really the honor should go to Jürgenson. Like his predecessor and mentor, Raudive found that the "messages from the dead" he captured were hopelessly inconsequential.

In a 1978 book, *Mediumship of the Tape Recorder*, D. J. Ellis published his analyses of the two men's work, coming to the conclusion that there wasn't anything of value there to analyze. Much later, in the 1990s, the psychologist Imants Barušs tried to replicate Raudive's experiments and came to roughly the same conclusion.

Another pioneer of EVP is the Italian medium Marcello Bacci. Building on the work of Jürgenson and Raudive, he initially used microphones and tape recorders, as they did, but eventually his instrument of choice for picking up EVP became an old Nordmende valve radio. He puts on monthly demonstrations, tuning his radio to a frequency somewhere in the 7–9MHz range, where commercial radio signals rarely stray, and then listens for whatever he might hear. After usually about ten to twenty minutes, the original background noise fades, followed by whooshing noises and then speech from one or more voices. These voices continue uninterrupted, even if Bacci twiddles the frequency knob. In late 2004, as an experiment, he tried removing the valves from the radio, and even that didn't perturb the spirit voices. According to the voices, such quirks are possible because they're communicating by means of "waves that are not physical."

The idea that spirits could speak to us through otherwise silent phone lines was popularized by the parapsychologist D. Scott Rogo in his book, appropriately named *Phone Calls from the Dead* (1979; with Raymond Bayless). Today, one of the foremost practitioners of the art of phoning the dead is Sonia Rinaldi, coordinator of the Brazilian Associaçao Nacional de Transcomunicadores (ANT).

Most of the people making use of her services are parents who've lost children. At an appointed time, they phone Rinaldi with a list of prepared questions. She has a phone hooked into her computer, so the conversation can be recorded, and an open extension phone for the spirits to speak into. As a background sound for the spirits to work with, Rinaldi plays three CDs simultaneously of "Portuguese-language crowd babble." The parents read their questions, leaving an interval of ten seconds between each one. Afterward, Rinaldi makes a copy of the recording and sends it to them. Apparently, the method works only when her callers are Brazilians living in Brazil; she has tried accepting international

EVP pioneer Marcello Bacci preferred to use an old Nordmende radio like this one as his detection device.

➤ PHONE CALLS FROM DEAD ANIMALS ❦

The Brazilian medium Sonia Rinaldi is among those who've been able to communicate via electronic means not just with dead humans but also with dead pets. In an article titled "ITC Contacts with Animals?" in the August 2008 issue of *Instrumental TransCommunication Journal,* Anabela Cardoso offers a roundup of various stories of human/animal ITC, including Rinaldi's own account of ITC contacts with a deceased parrot named Lorinho. Cardoso notes that she herself "heard some of the audio files containing the communications . . . and I can testify that I could easily understand most of the speech transcriptions that Sonia has sent me, together with the audio clips, pronounced in what sounded like a parrot's voice speaking in Portuguese."

Cardoso's own ITC contacts are with her "beloved deceased Doberman dogs"—and, indeed, dogs seem to be the deceased pets most often communicated with. One big communication advantage dead animals have over living ones is that they're apparently able, as implied above, to speak in the appropriate human language.

requests and when abroad has experimented to see what happens, but in both circumstances the results have been failures. Her hypothesis is that each nation has its own "station" on the other side.

WHITE NOISE AND THE SEASHELL EFFECT

So what are the possible natural causes of the "voices"? Leaving aside instances of fakery, part of the explanation is probably the auditory equivalent of pareidolia, the phenomenon whereby people see, for example, images of Jesus in pieces of toast, clouds, or dogs' bottoms.

When you play a blank tape, you're not actually listening to silence; there's invariably a soft hiss, even if you're not consciously aware of it. That subtle hiss is more correctly termed *white noise* and, when you listen to it, you can hear anything that you want to hear—old songs, children playing, the whispers of the dead, and so on. Similarly, dead phone lines usually have a hiss, and, of course, if you're holding a landline phone to your ear, you're likely to experience the additional hiss (and other sounds) typical of the seashell effect whereby someone holding a seashell to their ear can interpret the sound of ambient noise resonating in the shell's cavity—primarily the sound of their own blood flow—as the sound of ocean waves.

Furthermore, tape recorders don't operate in complete silence—they emit their little groans and creaks. If you're in a frame of mind to hear voices, it's easily possible to mistake one of these, at the very limits of audibility, for a word—especially since the white-noise effect may play a part. And, of course, if you listen to lots of odd little noises, chances are that every now and then one of them will come along that actually *does* sound like a word. Also, electronic equipment like tape recorders can sometimes pick up radio signals that they're not supposed to. And it's not impossible for a tape recorder to pick up a sound that the experimenters didn't hear because it was so quiet—a sound that, on replay, with the volume turned up to the max, can be quite distinctly audible. EVP enthusiasts are at pains to point out that they make great effort to eliminate all these possible causes—experimenting with equipment in radio-shielded rooms, for example. But, if the voices really do come from the afterlife, why don't any of the spirits have anything substantive to say? And, if they're from aliens, how come a civilization sophisticated enough to build a spaceship capable of traversing the distances between the stars can't communicate more effectively than by making faint noises on blank tapes?

Other researchers have pointed out that "spirit voices" find it easier to manifest themselves if there's a fair amount of white noise available to work with. Old-fashioned tape recorders yield better results than their virtually silent modern counterparts and a room with an electric fan in it is more favorable to EVP communications than a room that's noiseless. To most of us, these observations point toward the mundane explanations just outlined, but EVP researchers have an alternative interpretation: It seems the spirits can have only a very tiny direct physical influence on our world. They cannot manipulate the background sound itself, but they can make very subtle adjustments to the recording of it.

✦ THE PHYSICS OF THE SOUL

Writing during the first century BCE in *De Rerum Natura*, Lucretius speculated that the soul must be made of particles far tinier than those found in smoke, air, or water. The ancient Egyptians believed the soul was located in the heart, and by the time of Leonardo da Vinci, people instead thought it dwelled in the middle of the skull. As part of his anatomical investigations, Leonardo dissected brains, in search of the soul, and was accused of sorcery.

In 1854 the Austrian physiologist Rudolf Wagner attempted to persuade a congress of his colleagues at Göttingen to discuss the possibility that the soul was made of *stuff*—a "special soul substance." The congress allowed him to make his case, but afterward, in place of discussion, there was a stony silence. Had he waited just a few years until the Spiritualist movement really hit its stride, his suggestion might have fallen on more fertile ground.

SPIRIT PHOTOGRAPHY

Spirit photography arguably began in Boston in 1861, when a keen amateur photographer named William H. Mumler discovered, while developing the plate of a self-portrait, that there seemed to be an ethereal woman standing beside him. Mumler knew this was just an

The title page of a 1656 edition of *De Rerum Natura*, by the Roman philosopher Lucretius. In this work, written in the first century BCE, Lucretius speculated that the soul must be made up of particles far tinier than those in ordinary matter.

accidental double exposure. Friends, however, persuaded themselves that the woman was a deceased relative of his, and soon the newspapers were trumpeting the image as a photograph of the dead. Sensing a source of income, Mumler started producing spirit photographs at will—such as a portrait of Mary Todd Lincoln alongside the spirit of husband Abraham—using the very same method of double exposure. Unfortunately, people started to recognize some of the "spirits" as individuals who were still very much alive, and in 1869 Mumler was brought to trial for fraud.

Mumler was not an isolated example. Late in life, after the death of his beloved wife, Sir

One of Mumler's faked spirit photographs, showing Mr. John J. Glover with, purportedly, his mother's specter behind him, 1871.

William Crookes found solace in the spirit photographs of the English medium William Hope. One of the pictures Hope took showed Lady Crookes hovering at the shoulder of her husband; another bore a message for Sir Oliver Lodge. Lodge was unimpressed, pointing out that the spirit photograph of Lady Crookes looked remarkably like one that had been taken while she was still alive, at a wedding-anniversary celebration, and he asserted vehemently that the effect was just a double exposure. Crookes was equally convinced that the photograph was genuine—he had, after all, investigated the techniques of bogus spirit photographers back in the 1870s, when he'd been conducting his studies of the psychic.

A 1922 "spirit" photograph taken by Ada Deane, with crackpottery champion Sir Arthur Conan Doyle in the foreground.

Sometime before he died in 1919, he sent a print of the photograph to the Ghost Club with an inscription that he was doing so at the request of Lady Crookes.

A few years later, in 1922, Hope was tested at the College of Psychic Science in London by Harry Price of the SPR. Price was encouraged to bring his own plates, which, unknown to Hope, bore a mark that was visible only when they were developed. Standing beside Hope in the darkroom, Price quite clearly observed Hope swapping the plate he'd been given for one he'd stashed in his coat pocket. Sure enough, when the plate was developed, it lacked the identifying mark.

When Hope's fraud was exposed, there was an angry reaction from Sir Arthur Conan Doyle (page 57), who led a walk-out of members from the SPR and wrote a book defending Hope, *The Case for Spirit Photography* (1922)..

BARADUC'S FUZZY ORBS

In the early 1900s, the French physician Hippolyte Baraduc became intrigued by psychic photography. His belief was that the human body gave off radiation that was invisible to the human eye but could be detected by special photographic equipment. Initially, he envisioned this phenomenon as a potentially useful therapeutic aid. He was able to produce photographs showing, he claimed, what we might call auras indicative of people's moods. When he took photographs of the corpse of his son André, who died in 1907 of tuberculosis at the tender age of nineteen, one of the plates showed curious little speckles that—in Hippolyte's view—could only be something related to André's immortal soul.

A few months later, Baraduc's wife, Nadine, was dying—and this time he was ready with his photographic equipment to record the event. As she died, she let out "three gentle sighs" (to quote American psychic researcher Hereward Carrington). In his book *The Problems of Psychical Research* (1921), Baraduc described seeing three balls of light—and, sure enough, these balls appeared as little glowing blobs a few inches above her body in his photographs. He seemed—so he thought—to have photographed her soul as it left her body. Wrote Carrington:

> *These [luminous globes] gradually condensed and became more brilliant. Streaks of light, like fine threads, were also seen darting hither and thither. A quarter of an hour after the death of his wife, Dr. Baraduc took another photograph. Fluidic cords were seen to have developed, partly encircling these globes of light. At three o'clock in the afternoon, or an hour after her death, another photograph was taken. It will be seen from this photograph that the three globes of light have condensed and coalesced into one, obscuring the head of Madame Baraduc, and developing towards the right.*

American psychic investigator Hereward Carrington.

In the 1930s, Carrington, puzzling over the fact that a glowing ball had appeared as if

⇒ KIRLIAN CORONAS ⇐

A technique that for a while seemed to offer a gateway to physical investigation of the soul was Kirlian photography. In 1939, the Russian inventors Semyon and Valentina Kirlian discovered that if a living organism or piece of organic matter is photographed while in an intense electric field, the image shows a brightly sparkling "aura" around the subject. The discovery might have remained a curiosity in the history of science had it not been for its popularization in the best-selling book *Psychic Discoveries Behind the Iron Curtain* (1970) by Lynn Schroeder and Sheila Ostrander.

Many parapsychologists assumed the photographs showed some kind of "life field"—for which the term *soul* might serve just as well. In fact, what the photographs show is today recognized as coronal discharge, a perfectly natural phenomenon whose physics has become well understood.

from nowhere in a photograph of his own, satisfied to himself that he'd solved the mystery. The clue lay in the lighting setup arrayed behind the camera. Light sources to the back of the photographer were reflecting and scattering illumination from objects in front of him, creating the illusion on the photographic plate of glowing blobs. Another possibility, not suggested by Carrington, was that the fuzzy blobs of light could have come from tiny perforations in the camera's bellows.

WATTERS'S CLOUD CHAMBER

The invention of the cloud chamber in 1911 by the Scottish physicist C.T.R. Wilson inspired Hereward Carrington to suggest the same principle could be used to capture the image of a soul. In Wilson's cloud chamber, the air in an airtight container was saturated with water vapor and then expanded so that it cooled. Assuming the air to be completely dust-free, water particles would condense to form a "cloud" around the ions along the trail of any ionizing particle that should happen to be passing through the chamber. The invention earned Wilson a Nobel Prize in 1927 because of its usefulness in studying subatomic particles.

Carrington's version required an unfortunate animal to be killed inside a similar chamber, the water vapor–saturated air being expanded/cooled at the moment of the creature's death. Assuming the soul was made of ions, any resourceful researcher ought to be able to capture a photograph of it in the form of clouds condensing around those ions.

In the 1930s, the physicist R. A. Watters of the Reno-based William Bernard Johnston Foundation for Biophysical Research took up Carrington's idea using insects, frogs, and mice, which he decapitated in a cloud chamber with a cute little guillotine. Sure enough, misty shapes formed in the chamber following the moment of death. Moreover, Watters claimed, those misty shapes resembled those of the recently deceased animals. Discussing his experiments and their results in a chapbook called *The Intra-Atomic Quantity* (1933), Watters proposed his hypothesis that the ions of the soul occupied the spaces between the atoms of bodily cells.

Skeptical critics observed that the shapes generated in Watters's cloud chamber resembled the original creatures only in the same way that the shapes we see in the clouds of the sky sometimes make us think of animals—in other words, that Watters was a victim of a pareidolia effect. Later on, he improved his apparatus and methodology, and once again the enigmatic shapes appeared, as detailed in his *Cloud Chamber Investigations into Post-Mortem Ions* (1936). By this time, however, Watters had concluded that the ions in question weren't those of the soul but, more mundanely, the product of chemical decay.

Watters's experiments operated on the assumption that the soul left the body at the moment of death or very shortly thereafter, but the nineteenth-century Spiritualist Joseph O. Barrett—author of *Looking Beyond: A Souvenir of Love to the Bereft of Every Home* (1871)—had a different perspective. He was concerned about people being prematurely buried, not because they might be still alive but because the soul might not yet have had a chance to leave the body—a process he believed could take up to five days.

The Soul Hovering over the Body, Reluctantly Parting with Life, an illustration by William Blake for an 1808 edition of Robert Blair's poem *The Grave* (1743).

WEIGHING THE SOUL

At roughly the same time that the French physician Hippolyte Baraduc was attempting to photograph the soul (page 116), Dr. Duncan MacDougall of the Massachusetts General Hospital was attempting to weigh it.[5] He disagreed with Rudolf Wagner's assessment that the soul must be made of some special *stuff* unknown to the physics of the day, and likewise that it could be "etheric"— immaterial but still there. Instead, he believed that it must be made of mundane matter: solid, liquid, or gas. If so, he reasoned, the soul must have mass.

Trying to weigh the soul of a living person might be problematic— how do you separate the soul from the person? MacDougall reasoned that, if he could weigh a person's body immediately before and after death and find a difference, that difference must surely represent the weight of the departed soul. This was where his position as a hospital doctor came in handy. He recruited some terminal patients willing to help him with his research and, as their time drew near, set up monitoring equipment around them.

He found that patients steadily lost small amounts of weight in the hours leading up to death and attributed this to moisture loss through respiration and perspiration. But this didn't explain the fact that his instruments recorded a sudden small weight loss—about three-quarters of an ounce (21 g)[6]—at the moment of death, as well as a small weight loss a few minutes later. When he expanded his experiments to include dogs, he found no such weight loss; clear evidence, he concluded, that only humans have souls.

Nervous about incurring public ridicule, MacDougall held off

---

[5] MacDougall was far from the first to think the soul might have weight. The ancient Egyptians had a similar notion, which is why tomb paintings sometimes show the god Anubis weighing the soul of the deceased, using a balance with a feather in the other pan.

[6] Hence the title of Alejandro González Iñárritu's neo-noir movie *21 Grams* (2003).

SOUL HAS WEIGHT, PHYSICIAN THINKS

Dr. Macdougall of Haverhill Tells
of Experiments at
Death.

LOSS TO BODY RECORDED

Scales Showed an Ounce Gone in One
Case, He Says—Four Other
Doctors Present.

Special to The New York Times.

BOSTON, March 10.—That the human
soul has a definite weight, which can be
determined when it passes from the body,
is the belief of Dr. Duncan Macdougall, a
reputable physician of Haverhill. He is
at the head of a Research Society which
for six years has been experimenting in
this field. With him, he says, have been
associated four other physicians.

The *New York Times* article that
blew MacDougall's cover.

publishing his findings for about five
years. Eventually, the appearance in
March 1907 of a sensationalized report
on his experiment in the *New York Times*
spurred him to publish his "Hypothesis
Concerning Soul Substance Together
with Experimental Evidence of the
Existence of Such Substance" a few
weeks later in both *American Medicine*
and the *Journal of the American Society
for Psychical Research.*

The scientific community didn't
take a lot of notice of MacDougall's
findings. For one thing, his measuring
equipment was judged to be a bit Rube
Goldberg. Second, MacDougall had
worked with just six dying patients and
had discounted a couple of the results as
being mere outliers—an alarmingly high discard rate. (He himself
was wary of the low sample size, noting, "I am well aware that these
few experiments do not prove the matter any more than a few swal-
lows make a summer.") Another Massachusetts physician, Augustus
P. Clarke, pointed out that, at the time of death, there's a sudden
increase of body temperature because the blood is no longer being
cooled by circulation through the lungs; as a result, you could expect
a peak in the evaporation of sweat, with consequent weight loss.
However, this explanation seems to have its problems, too, in that it's
unlikely that as much as twenty-one grams of sweat could evaporate
instantaneously. It would, however, explain the disparity of the results
between humans and dogs, in that dogs don't sweat.[7]

---

[7] They all go to heaven, though!

In his self-published book *The Physical Theory of the Soul* (1915), physicist Harry LaVerne Twining recounted his experiments to determine if mice underwent a similar weight loss at the time of death. His early experiments, in which mice were killed using a button of cyanide, did indeed show a loss of weight upon the creatures' expiration. The loss was tiny—about one to two milligrams—but not zero. "An unscientific person or a philosopher," he asserted, "would rest satisfied and conclude that the mouse had

Augustus P. Clarke, a perhaps overenthusiastic debunker of MacDougall's work.

a soul, and at the moment of death, he would conclude that it left the body, thus causing the loss of weight. He would also conclude that the soul had weight."

But Twining himself was made of sterner stuff. He repeated his experiment several times more, stuffing the mice into test tubes whose open ends had been sealed off, so that the unfortunate creatures smothered to death.

> *Under these conditions, no loss of weight took place whatever when the mouse died. The dead mouse and tubes were left on the scales over night but no change was observed in the morning. Whatever the mouse lost on dying then could not get out of the glass tube. If this were the soul it would be rather an unfortunate predicament for it. A human soul sealed up in the air tight casket would have to rest confined until the casket rotted or disintegrated before it could escape.*

He then experimented using various other means of killing the mice and was able to show that, despite the conclusion of Baraduc and others in the case of humans, the decline in weight was due to loss of moisture. Twining might have concluded that this meant that mice didn't have souls, but instead he harked back to Lucretius: "The material of the soul does not have weight because it is finer than the material that causes weight in coarse matter. The soul must have mass, however, although it has no weight. Weight is not a property of matter but mass is."

CAPTURING SOUL-STUFF

The German zoologist Gustav Jäger (or Jaeger) was, in effect, "Darwin's Bulldog" in Germany. He also promoted an unorthodox theory of heredity, later elaborated by others, that the germ plasm divides during reproduction, with one part being dedicated to the development of the individual offspring and the other being reserved for later, to be transmitted when the offspring reproduces. But he achieved arguably his greatest historical fame as the creator of Jaeger Sanitary Woolens, a line of underwear and, later, outerwear.

The inspiration behind his development of the clothing was his finding, reported in *Die Entdeckung der Seele* (1878; translated as *The Discovery of the Soul*),

German naturalist Gustav Jäger, shown in this 1884 photo wearing his "sanitary woolens," investigated, among much else, the odor of the soul.

that the soul has an odor that you can actually smell as it leaks out from your mouth, nose, and skin. Through its smell, Jäger was able to trace, by experiment, the seat of the soul to the brain. He pulverized a small quantity of brain matter in a mortar, added nitric acid, and thereby produced the perfume of "soul-stuff."

Researching further, he found that the soul wasn't the only intangible entity with a smell; emotions had them, too. Collectively, he termed these odoriferous mental/spiritual aspects *dufts* (German for "fragrances"). If a *duft* smelled nice, it was beneficial; if it smelled bad—for example, the jealousy *duft*—it was nasty. Ideally, then, one would like to retain the beneficial *dufts* around oneself while letting the negative *dufts* dissipate. Luckily, Jäger discovered there was one material with the astonishing property of absorbing positive *dufts* and expelling negative ones: wool. For the optimum soulish health, therefore, his prescription was to wear nothing but woolen garments right against the skin.

The wool had to be 100 percent pure. However, there was no reason to adhere blindly to sheep wool: The wool of other animals could supposedly work just as well and perhaps would not be so infernally itchy around the hindquarters. In the end, he settled on camel wool as his ideal material. An additional interesting property of camel wool was that, if you wore nothing else by day and by night, your desire for food dropped off markedly—so Jaeger Sanitary Woolens became a diet aid, too.

The hair was the best place to search for a reservoir of an individual's *duft*. Jäger demonstrated that, after several deep sniffs of his youthful soul-stuff (as preserved in a lock of his own hair, which he'd given his wife as a love token decades earlier), his reaction times improved markedly. Similarly, by preparing a solution of wine and *duft* from a lock of a lovely young woman's hair, you could drink nearly twice as much before falling over, even as your good humor would keep increasing in the desired fashion. Furthermore, your

singing voice would improve in both tone and range. This finding—
that food and drink could be improved by being "humanized"—was
put to commercial use. Jäger marketed the essences as "anthropines,"
which were numbered to distinguish the fragrance for, say, curing
arthritis from the fragrance intended to invigorate the body.

Not all the attempts to explain the scientific nature of the soul
have been so, well, grounded. In *Human Personality and Its
Survival of Bodily Death* (1903), F.H.W. Myers offered this
observation:

> *I claim, in fact, that the ancient hypothesis of an indwelling
> soul, possessing and using the body as a whole, yet bearing a
> real, though obscure relation to the various more or less
> apparently disparate conscious groupings manifested in
> connection with the organism and in connection with more or
> less localized groups of nerve-matter, is a hypothesis not more
> perplexing, not more cumbrous, than any other hypothesis yet
> suggested. I claim also that it is conceivably provable,—I
> myself hold it as actually proved,—by direct observation. I
> hold that certain manifestations of central individualities,
> associated now or formerly with certain definite organisms,
> have been observed in operation apart from those organisms,
> both while the organisms were still living, and after they had
> decayed. Whether or not this thesis be as yet sufficiently proved,
> it is at least at variance with no scientific principle nor
> established fact whatever; and it is of a nature which continued
> observation may conceivably establish to the satisfaction of all.*

Basically, he dismisses those who claim there's no such thing as a soul
by arguing that it's impossible to prove that there isn't.

In *The Seat of the Soul* (1989), Gary Zukav tells us, "The individual unit of evolution is the soul," and expands hither and thither

about the physics of the immortal soul—some parts of which I confess I still find hard to understand, even after Zukav has told us that they're not hard to understand at all.

It seems that souls are eternal—they have "no beginning and no end"—but at the same time some of them are older than others. Zukav is at pains to reassure us that there's no paradox here; it only seems paradoxical because in our ordinary way of thinking we assume that things must have a beginning. If things don't have a beginning— if they've always existed—then different rules apply. In trying to understand this, I thought of the analogy of the transfinite numbers. If you start counting the positive integers—1, 2, 3, 4 . . . — you can obviously keep on going forever: there are an infinite number of them. Now consider adding to your heap of positive integers all the negative ones: −1, −2, −3, −4 . . . Plainly, assuming you'd stopped before you got to infinity, you'd now have a heap that contained twice as many numbers as your first one. If you then start thinking about the number of different orders in which you could arrange all those numbers, you get something even bigger. Take this to infinity and . . . and then you get stuck, because there's no sense in talking about one infinity being bigger than another—they're all infinity.

I felt I was beginning to get some sort of conceptual handle on the notion of some eternal souls being older than others, but then I read, "All that is can form itself into individual droplets of consciousness. Because you are part of all that is, you have literally always been, yet there was the instant when that individual energy current that is you was formed," and I was lost again.

3

TO DESTINATIONS UNKNOWN

TRAVELS TO
THE GREAT BEYOND . . .
AND BACK!

"I feel the presence of my dog around me as I ponder those two
questions. Then I hear barking, and other dogs appear, dogs I once
had. As I stand there for what seems to be an eternity, I want to
embrace and be absorbed and merge. I want to stay. The sensation
of not wanting to come back is overwhelming."

—Bryce Bond, cited by P.M.H. Atwater in *Beyond the Light: What
Isn't Being Said About Near-Death Experience* (1994)

THE TERM *OUT-OF-BODY EXPERIENCE*[1] WAS
coined by the mathematician G.N.M. Tyrrell, who was
president of the SPR in 1945–1946. He proposed it in his
Frederic W. H. Myers Memorial Lecture to the Society
in 1942; an expanded version of that lecture appeared the following
year as *Apparitions* (1943). Tyrrell wanted to replace terms like *astral
projection* that implied a supernatural element. Many people have
experienced one or more OOBEs, perhaps through exhaustion, stress,
booze, anesthesia or drugs (legal or otherwise), or perhaps for none
of these reasons. Some people can induce OOBEs at will.

One of the largest surveys of OOBEs was carried out by the
British philosopher and parapsychologist Celia Green at the Institute
of Psychophysical Research (IPR) in Oxford. She gathered together
and analyzed four hundred OOBE accounts, attempting to classify
them into categories. She linked OOBEs to lucid dreaming[2] and the

---

[1] There are various common abbreviations for the term: OOBE, OBE, or even
OOB. I prefer OOBE.

[2] Dreaming in which the sleeper is aware that she or he is experiencing a dream.
Some people do this frequently (and can even start controlling the events of the
dream); others never do. I have had exactly one lucid dream in my life. It occurred,
oddly enough, while I was writing this book.

seeing of apparitions, proposing that in all cases our set of normal senses is taken over by a hallucinatory one. She was, therefore, postulating an entirely nonsupernatural mechanism for OOBEs, and today most psychologists (and, indeed, parapsychologists) agree with her on this, although not necessarily with her suggested mechanism.

In simple terms, there seem to be three basic forms of OOBE. In the first, the "observer" never travels far; hospital patients find themselves seemingly floating above the operating table, looking down at their own bodies, for example. Something similar happens to all of us in a small way quite frequently, as when we suddenly find we're hearing ourselves talking as if we were someone else.

In the second kind of OOBE, the "astral projection" travels far from the body and may possibly, according to anecdote, be seen as an apparition by people in that distant place. Perhaps related in some way to both these forms of OOBE is the doppelgänger effect, wherein individuals see an apparition of themselves.

The third distinct category of OOBE is the near-death experience (NDE), in which experients typically believe they've gone to heaven. While psychologists and most parapsychologists readily identify these experiences as OOBEs and therefore nonsupernatural, there's a burgeoning NDE industry that would have you believe otherwise: Those trips to heaven must be real, and the accounts of the happy (and often

Samuel Taylor Coleridge was transported to Xanadu during an opium-induced nap in 1797. What he reported in his poem "Kubla Khan" about this heavenly paradise was remarkably vivid! Perhaps most OOBEs and NDEs have a similar hallucinatory origin?

First published in Camille Flammarion's *L'Atmosphère: Météorologie Populaire* (1888), this engraving shows a man crawling under the edge of the sky to look at the empyrean beyond. Modern visitors to heaven have taken a different approach.

thereafter wealthy) survivors must be confirmation that there is, indeed, life after death. Although the NDE is the form of OOBE that most concerns us in this book, it makes sense to first consider OOBEs in general.

Also seemingly related to the OOBE is the purported phenomenon of "remote viewing." It's quite difficult to figure out what the difference is between remote viewing and the purported ESP faculty called clairvoyance, but it was precisely to make that distinction that Harold Puthoff and Russell Targ of the Stanford Research Institute[3] coined the term. As far as I can understand it, the difference is that

---

[3] Which has nothing to do with Stanford University, except that it is in a nearby geographical location—a point often overlooked in reports of the SRI's activities.

the clairvoyant sees the distant place/event without actually "going" there, while the remote viewer in some way travels out of the body in order to see what's going on at the remote location. Between 1975 and 1995, the US government invested quite a lot of taxpayer dollars in remote-viewing research on the grounds that it might prove a useful means of intelligence-gathering. As is so often the case, the two meanings of the word *intelligence* seem at odds.

❧ ASTRAL ADVENTURES . . . OR INVENTIONS?

In an experiment that he reported in 1974[4], ASPR researcher Karlis Osis asked about a hundred people who were convinced they could induce OOBEs at will to OOBE their way to a picture or objects that he had placed in another room and then, on their return, to describe what they had seen. Despite most being confident that they had done as asked, their descriptions of the room's contents were essentially random noise. Clearly, while the OOBE had been very real to them, there was no objective evidence that they had traveled anywhere.

This came as a disappointment to Osis because earlier experimentation that he had done in 1971 with the medium Ingo Swann had seemed to indicate that perhaps Swann was capable of OOBEing. The results weren't dramatic, but they were at least interesting. The same could be said of experiments Osis did in 1974 with the psychic Alex Tanous. In the latter instance, Osis took considerable steps to differentiate between what Tanous might see via OOBE and what he might see through the actions of ESP. He also analyzed very thoroughly the data from all the trials he ran. The risk in such rigorous analysis is that you may

---

[4] W. G. Roll, R. L. Morris, and J. D. Morris, eds., "Perspectives for Out-of-Body Research," *Research in Parapsychology, 1973* (1974).

⇒ A CLASSIC HOSPITAL OOBE ⇐

The classic form of OOBE—at least in folklore—is the one that happens to a hospital patient during surgery. After recovering from anesthesia, the patient will recount having floated up to the ceiling, watched from above as the surgeon and nurses went about their work, seen what someone once left on top of a cupboard, and so on. It's often treated as among the remarkable aspects of a hospital OOBE that the patient can later describe what the surgeons and nurses were talking about during the procedure. In fact, this is the least remarkable aspect of all. Hearing is the sense that's least susceptible to anesthesia, and it's not especially uncommon for people to come out from "under" with at least some recollection of what they've heard while they were apparently completely unconscious.

A classic hospital OOBE involves a woman named Maria who, in April 1977, was rushed to Seattle's Harborview Medical Center following a heart attack. A couple of days after she was admitted, she suffered a short period of cardiac arrest. Later she told how, while the staff labored to save her, she drifted from her body and then from the room, giving herself a little tour around the outside of the hospital. Most of what she described was not puzzling—things she could have seen while arriving at the hospital but, understandably under the circumstances, had not

eventually find something statistically significant, even if in fact it's not really meaningful. For example, the results of the experiment with people who tried to OOBE their way to a distant room and identify objects therein suggested nothing more than chance at work. A deeper analysis, however, might have revealed they did better with pictures of teddy bears than they did with pictures of tapeworms. Is that fact significant? Or just probability (or even psychology) at work?

registered at the time. One item stood out, however. Perched on a third-floor window ledge, she reported, there was an old tennis shoe with its laces arranged in a particular way. The counselor to whom she explained all this, Kimberly Clark,[1] eventually discovered just such a shoe, exactly where Maria had said she'd seen it.

For some reason, it was the mid-1980s before Clark published the story, and the mid-1990s before a team of skeptics decided to check it out. They put a tennis shoe on the ledge in question and discovered that, contrary to Clark's account of having to put her face against the glass and squint before she could see it, they could do so quite easily from inside the room—even from the vantage point of a patient in bed. Furthermore, the shoe was also clearly visible from the ground outside. While of course this doesn't conclusively disprove Maria's OOBE, it seems much more probable that Maria learned about the shoe by, for example, hearing a couple of nurses chatting about the oddity of a shoe on the hospital's third-floor window ledge, and that, in the years between the event and Clark writing her account, the tale was unwittingly embroidered a little. We don't know that things happened exactly like this, but it's a perfectly reasonable explanation that has no need of the supernatural.

---

[1] Now Kimberly Clark Sharp, she is regarded as an expert on OOBEs.

In her book *Beyond the Body* (1982), Susan Blackmore describes the experiences of two people who, during the early decades of the twentieth century, were able frequently to project themselves astrally, or at least have an OOBE that seemed to them to be astral projection: Oliver Fox and Sylvan Muldoon. Both gave extensive accounts of their astral adventures. Reading Blackmore's summaries of these, however, it's hard to distinguish their experiences (and those of later astral travelers whom Blackmore

⇒ TRIPPING ⇐

Blackmore's own interest in OOBEs was kindled when, in the early 1970s, as a student at Oxford University, she herself underwent an OOBE through a combination of exhaustion and marijuana. In common parlance, she was tripping. Although it's unusual to trip on marijuana rather than a hallucinogen like LSD, it's not unheard of, especially when the person is exhausted. (Ditto for alcohol.) The relationship between OOBEing and tripping would seem to be close—close enough, in fact, that it's tempting to regard the OOBE as a special case of tripping.

Blackmore clearly experienced floating up to the ceiling, then over brick-red chimneyed rooftops. She also saw the fabled silver cord joining her physical body to her "astral" one, reached out a hand to move the cord, and "immediately learned my first lesson. I needed no hand to move the cord, thinking it moved was sufficient. Also I could have two hands, any number of hands, or no hands at all, as I chose." This discovery, even before she found out the following morning that the roofs over which she had "flown" were neither red nor chimneyed, was enough to convince her that at least this particular OOBE was no physical venture beyond the body . . . unless, as others have theorized, the world through which our astral bodies travel is an inexact duplicate of our own.

discusses) from various forms of lucid dreaming. As Blackmore notes in one case:

As with so many other OBEs, the details seen tended to be a mixture of right and wrong; enough right to make one feel that more than chance is involved, and enough wrong to be sure that the OBEer is not seeing a complete duplicate of the physical world at that time.

Blackmore herself concludes there's no compelling reason to believe that any OOBE represents a paranormal event. In fact, by her own definition of the term OOBE—that it should refer to the experience, rather than a purported trip outside the body—the paranormal hardly comes into play. After all, the experience of seeming to travel out of the body is genuine enough for the experient, even if no such trip takes place. Her attitude is that we can study OOBEs without having to take the paranormal into consideration unless it forces itself upon our attention.

CHRISTOS

The Australian novelist G. M. Glaskin first came across a reference to the Christos Experiment (or the Christos Experience) in 1971 when he saw a magazine article about it. The technique for "astral travel to previous lives" had been devised not long before by "researchers" Nicolas and Jacqueline Parkhurst.

The procedure began with the subject lying down with shoes off and the head supported on a pillow. One of two attendant aides then rubbed the subject's pineal region[5] vigorously with the inside edge of a clenched fist while the other massaged the subject's ankles.[6] After a

---

[5] The region of the forehead where the pineal eye would be if humans had a pineal eye. The belief that we actually do have one pops up in various schools of mysticism; it's the "third eye" that the British mystic T. Lobsang Rampa made so famous from the mid-1950s onward.

[6] Glaskin himself eventually discovered that he could induce the experience without assistants.

One of the more unusual contraptions dreamed up to give humankind the power of flight, from *Wonderful Balloon Ascents* (1870) by "Fulgence Marion"—i.e., Camille Flammarion. Glaskin's method purportedly offers a similar experience without all those pesky ropes and appendages.

while, the person in charge of this particular run (the "runner") would ask the subject to visualize himself or herself growing taller, first by a couple of inches (roughly 5 cm) from the feet and then by the same amount in the opposite direction—that is, from the head. Then the length was increased a bit, and so on, until he or she had "grown" some twenty inches (50 cm or so) in both directions. Next, the subject was instructed to expand like a balloon and "travel" to the front door. After further exercises, the subject was told to zoom off in a random direction.

Glaskin and various friends experimented at length with the technique, and he described the results in a series of books, beginning with *Windows of the Mind* (1974). It's obvious from reading these books that the actual experience of these "artificially induced OOBEs" (assuming that's what they were) was quite startling. The subjects saw everything in incredible detail and with great intensity. Throughout it all, they were conscious in both "worlds"; that is, they could hear and react to the runner's instructions even as they explored "elsewhere."

That the "journeys" were more likely a product of the subjects' own minds than genuine external perceptions is clear from the fanciful nature of some of the observations. In one of Glaskin's own "runs," for example, he found himself in the body of a prespeech hominid. He fled from his own people through a long tunnel, only to be welcomed by a more advanced group who started to teach him the rudiments of speech. The oddity here was that, though the scene was set many thousands of years ago, the tunnel was lit by electricity, if not a more advanced technology than that. "Again," Glaskin elucidates, "I had a kind of precognition that it was the accomplishment of a prior and much more civilized race that was superior even to twentieth-century man, moon-landings and all notwithstanding."

Glaskin thought that the subjects were becoming "consciously receptive to [their] subconscious without the normal requisite of sleep"; in a sense, then, the runs were extraordinarily vivid waking dreams. Reading through the accounts of the Christos runs in Glaskin's books, it's noticeable how many of them seem either to have taken the form of wish-fulfillment fantasies or to have helped the subjects sort out a problem that was troubling them in their waking lives. It seems thunderingly obvious that the Christos Experience offers no evidence for the existence of an immortal soul, but perhaps it has psychotherapeutic possibilities.

PRACTICAL PROBLEMS

A question that flummoxed many researchers—such as Eleanor Balfour (Mrs. Henry Sidgwick) when researching apparitions for the SPR in the 1880s—is why people traveling out of their bodies are, when seen by others, generally fully clothed. Do our pants and socks possess astral selves in the same way that we do? It's a problem that seems never to have been fully resolved.

⇒ AUTOSCOPY ⇐

The sensation of seeing yourself as if from outside is recognized in psychology. The term for it is *autoscopic hallucination*. In order to understand how it might work, we have to distinguish between your brain, which is your organ of cognition (and much else), and your own self—which, for ease, we'll call the mind. While you know where your brain is, your mind doesn't have any such precise location—or, really, any physical location at all. However, because your primary means of perception is your eyes, followed by other senses whose organs are mainly clustered around your skull, you tend to assume that your brain and your mind are in the same place.

It's easy enough, though, to persuade the mind that it's located elsewhere. Hallucinogenic drugs can do it, as can fever, a brain tumor, dementia, and so on—even a surfeit of liquor. In a Swiss experiment reported in 2007, researchers fitted subjects with virtual-reality goggles set to show the images of a camera placed behind them, so that it looked to them as if they were staring at their own backs. It was surprisingly easy to convince their brains that this was exactly what they were doing, and that the mind was observing from somewhere outside the body.

Another troubling problem that has to be confronted by those who maintain that the OOBE is a physical event, rather than a psychological one, is the question of vision. If the "astral body" is invisible and otherwise physically undetectable, light waves should pass right through it. How then could the eyes register visual information? The counterargument is that perhaps the astral body detects the world around it using something other than light and is able to do this without some sort of analogue of the eye. This boils down

Autoscopic hallucinations may explain at least some reported doppelgänger events—instances when people believe they've encountered their spectral twin. A more powerful hallucinatory explanation for doppelgängers is the "subjective doubles syndrome," in which sufferers see someone else—a friend, a relative, a neighbor, a stranger—not as the person really is but as a physical duplicate of the sufferer.

A related notion, proposed by parapsychologists Susan Blackmore, John Palmer, and others, is that, during periods when the brain is receiving little by way of sensory information—such as in the hypnagogic state (page 102)—it has a tendency to fill in the gaps with spurious information. Examples of such periods come when we're unconscious, sleeping, or knocked out by anesthetics in the operating room. In Palmer's psychoanalytic model, the lack of direct sensory information threatens a person's sense of individual identity, prompting the unconscious mind to reach around for some other way to define itself; in simplistic terms, it begins to fantasize. The unconscious mind then persuades the ego that the new, fanciful world it has conjured up is reality.

One speculation put forward by parapsychologist Michael Grosso is that perhaps our mind is *always* located outside the body. In this view, OOBEs represent those occasions when we simply become aware of that fact!

fairly quickly to a model in which the perception is being registered directly by the mind. But if *that's* possible, there's no necessity for any part of us to go traveling at all—using direct mental perception, we could just as easily "see" all the sights without moving from our beds. In other words, we're back to ideas of the OOBE as a mental—not a physical—event.

The idea of the astral body being connected to the physical one by a thin silvery thread is venerable. Plutarch, in his treatise *On the*

A portrait of Plutarch, who wrote an early account of an NDE in the first century CE.

Delay of Divine Justice, wrote about a 79 CE episode concerning a degenerate named Aridaeus of Soli, who was a friend of Plutarch's friend Protogenes. Knocked unconscious by a fall and for three days believed to be dead, Aridaeus went to Hades, where he met up with an uncle who pointed out to him that, since his thread was still intact, he couldn't really be dead. He "returned to life" just as he was about to be buried, and thereafter became an exemplary character.

The silver cord is noticeable by its relative absence from the records of medical scientists who've investigated OOBEs and NDEs. This emphasizes the point that what people discover during their OOBEs and NDEs seems to be dictated by their expectations. As we'll see, the same goes for the spiritual authority figure to whom many NDEers must render a "life review": Christians meet Christ, Muslims meet the Prophet, and so on. Interestingly, even atheists and agnostics are likely to encounter such an incandescent figure, which could be taken to mean either that, in the hinterland between life and death, we really *do* meet up with a transcendent authority or that the experience is a quasi-rationalized, narrativized version of the "life flashing before my eyes" phenomenon that seems to be a by-product of our memory mechanisms (page 151).

❧ DRIFTING TOWARD THE LIGHT

Before plunging into the realm of NDEs, visits to the afterlife, and the like, it's worth noting that medical science now recognizes several different definitions of death. The condition of *brain death* means—as we'd expect from the name—that brain function has ceased entirely; the dead brain and central nervous system can no longer perform even the autonomic functions like breathing and heartbeat. Then there's *biological death*—usually concomitant upon brain death—in which the various organs of the organism have irreversibly shut down. The important condition in the context of NDEs is *clinical death*, which is different from the other two in that, while the body may seem to have shut down, the brain and organs remain in such a state that there's still a chance of reviving the patient. In other words, clinical death is a condition from which there's a definite hope of recovery. However, because the term seems so final and, well, clinical, most of us tend to assume that a recovery from clinical death is more remarkable than one from, say, brain death. This misperception has been much exploited by self-publicists who'd like you to believe their recovery from clinical death is something miraculous.

LIFE AFTER LIFE

One of the smash hit best sellers of 1975 was *Life after Life* by a one-time philosophy professor named Raymond A. Moody Jr.

Moody, who coined the term *near-death experience*, interviewed hundreds of people who had suffered a brief period of clinical death before being resuscitated. From these interviews he was able, he stated, to observe an astonishing number of parallels between all NDEs—no matter what the cultural or religious background of the experient. Near the start of the book, he offers a description of, as it were, the "ideal" NDE—one containing all the frequently observed attributes of these experiences. In practice, he stresses, any one particular NDE is unlikely to go through every single one of the stages he outlines, but it will certainly display some and quite likely most of them.

Several of these claimed characteristics of NDEs have been sufficiently absorbed into Western culture to attain the status of "recognized scientific facts," even among people who do not share Moody's beliefs about afterlife survival (the book's subtitle is *The Investigation of a Phenomenon—Survival of Bodily Death*).

Moody says there are "about fifteen separate elements"[7] involved—that is to say, fifteen stages through which a typical NDE is likely to, but need not necessarily, go. In *Reflections on Life after Life* (1977), Moody adds a few more shared elements that emerged during further interviews with NDE survivors. These "are far from being as common as the original fifteen."

In Moody's reporting on NDEs, the most otherworldly element would seem to be the interaction with the "being of light." In *Life after Life*, Moody explains that his interviewees differ along broadly religious lines as to the identity of this being, but only broadly; for example, not all Christians assume the being is Christ or an angel. Even so, Moody draws a parallel between descriptions by NDE experients of the "being of light" and the description given by Paul in Acts 9 of his encounter with God on the road to Damascus. However, Paul specifies in Acts that the voice spoke to him in Hebrew, whereas Moody's interviewees said the light didn't use a language or speech at all, but communicated with them through a sort of transmission of understanding. So we're down to just a supposed parallel between two bright lights that have a message to convey.

---

[7] Working from his description, in *Life After Life*, I could come up with only twelve. I quickly checked his entry in Wikipedia to see where I'd gone wrong, and discovered the editors and contributors there had been able to discern only nine. This made me feel less inadequate!

Detail from an anonymous sixteenth-century painting of the conversion of St. Paul. Did the Evangelist have an NDE?

In a much later book, *The Last Laugh* (1999), Moody is at pains to point out that he doesn't believe NDEs in themselves constitute proof of life after death. At the same time, he obviously does believe in life after death for other reasons because, through hypnotic regression, he has accessed at least nine of his own previous lives. He has also founded in Alabama an institution called the Doctor John Dee Memorial Theater of the Mind, "devoted to the use of altered states of

John Dee and Edward Kelley attempting to raise a spirit in a churchyard. Dee believed Kelley had the ability to contact angelic beings.

consciousness for the purposes of education, entertainment, and spiritual advancement." Those altered states of consciousness are attained through scrying (techniques such as crystal-gazing), and the idea is that, via scrying, you should be able to see apparitions of the dead. This was, indeed, the method used by the sixteenth-century mathematician and occult philosopher Dr. John Dee and his rapscallion colleague Edward Kelley in order to, as Dee believed, communicate with angelic beings.

DEATHBED VISIONS

Moody was far from the first to investigate NDEs. In his *Looking Beyond* (1871), the Spiritualist Joseph O. Barrett gives a number of accounts of NDEs and similar experiences. Here's one, written by someone named J. M. Peebles about a recently deceased acquaintance, the Reverend J. W. Baily of Lima, New York. Baily's widow wrote to Peebles after her husband died, and Peebles retold this to Barrett:

> *The day before he passed he began to sing, and would sing for hours. Mrs. Baily asked him, "Does it not tire you to sing so much?" "O, yes," said he; "but I'm so happy—happy, I can't help it."* . . . *[Mrs. Baily] says he then turned his eyes upward, and ["]O, how glorious they looked! They seemed illumined with heavenly light; but he stopped breathing. I laid my hand upon his shoulder. He opened his eyes, and smiled on me, and said, 'Why, I thought I had gone to the spirit world. I have seen over the river, and I can now see on both sides. It is beautiful on this side; but O, glorious, glorious on the other!* . . . *I see so many friends there, over the river, and they beckon, beckon to me. I see more, vastly more on that side than I do on this.'" Mrs. Baily adds, "He then pressed my hand, said 'do not grieve,' smiled, waved his hand, and passed on."*

The real pioneer in the field, though, was another Barrett—Sir William Barrett, who published the results of his researches as *Death-Bed Visions* (1926). Most of Sir William Barrett's cases involved the patient being greeted by loved ones who had died earlier and who seemed to have come in order to assist the individual's transition to the afterlife.[8]

The monograph *Deathbed Observations by Physicians and Nurses* (1961) by Karlis Osis went into all this in much more exhaustive detail,

---

[8] In several books, Colin Wilson claims that the welcomers always represent the dead, that never does the patient encounter the "souls" of those still living, and that there are cases on record of patients being met by loved ones whom they did not know had died. I've been unable to evaluate the validity of this claim.

hoping to sort out—through analysis of things like religious beliefs and cultural differences—if the patients were genuinely seeing what they thought they saw or if their near-death visions were merely expressions of commonly held, culturally determined comfort fantasies. The study was based on the results of ten thousand questionnaires sent out to hospital doctors, nurses, and primary physicians; unfortunately, he received only 640 responses. From those responses, about one-quarter of which were followed up in more detail by mail or telephone, Osis was able to derive records of over 1,300 accounts of people seeing apparitions as they died, nearly 900 who'd seen idyllic scenery, and about 750 who were reported as becoming increasingly blissful.

An expanded research project, based on further surveys—of nearly two thousand doctors and nurses, all told, in the United States and in northern India—was published as *At the Hour of Death* (1977) by Osis and Erlendur Haraldsson. Although Christian, Hindu, Muslim, and Jewish patients reported differing NDEs (with the Christian and Hindu patients, in particular, showing the biggest contrasts), the researchers chose to view their results as indicative that, indeed, "the survival model" was correct.

In part, they drew this conclusion because of the nature of the figures whom the experients encountered during their deathbed visions. All told, about 80 percent of those visionary encounters were with either religious figures or dead people—usually dead loved ones. (In ordinary visions, the ratio is far lower.) Factors such as age, sex, and any drugs that the patient was receiving had little or no effect. One cultural difference was that about one-third of the Indian patients resisted what they saw as forceful persuasion by the visionary figures to join them in the afterlife, whereas the US patients were by and large likely to be willing to make what they clearly experienced as a consensual and peaceful transition. One interesting statistic is that, out of all the recorded deathbed visions, only one was a vision of hell! Clearly we're almost all optimists as we die.

The trouble with this project was that it relied on accounts given to Osis and Haraldsson by doctors and nurses of what very ill patients had told them, sometimes long ago—hearsay evidence, in other words. There are all sorts of ways in which the information could have been corrupted. As with any study, there's also the matter of possible unconscious researcher bias. Osis apparently became interested in the psychic through an experience he had as a teenager growing up in Latvia. (At the moment an aunt died elsewhere in the family home, the young man felt a sudden wave of delight pass through him—in direct contrast to the sorrow of the occasion. Could he have somehow received an overspill of the aunt's emotions on leaving behind this vale of tears?) After gaining a doctoral degree in psychology from the University of Munich for a study of ESP, he came to the United States and worked for a while with parapsychologist J. B. Rhine at Duke University before becoming research director at the Parapsychology Foundation. By the time of these two surveys, he needed very little pushing to be "persuaded" of the "survival hypothesis." Similarly, Haraldsson had spent time at the Institute for Parapsychology in Durham, North Carolina, and at the ASPR, and he would later work with Rhine and reincarnation proponent Ian Stevenson (page 176). In short, both men likely had favorable preconceptions when it came to the matter of survival after death.

AND WHAT OF THE SCIENCE?

Various researchers have raised fundamental questions about Moody's reported results. One big problem is that he has steadfastly refused to reveal to those wishing to replicate his work the identities of his interviewees, claiming that he wishes to safeguard their privacy. This might seem reasonable enough, but any ordinary scientist would ask the interviewees themselves whether or not they wanted to cooperate with other researchers. It is puzzling that, out

Hieronymus Bosch's painting *Ascent of the Blessed* (c. 1500) uncannily evokes modern accounts of NDEs.

of all the hundreds of interviewees Moody has claimed, not one has stepped forward.

In *Reincarnation: A Critical Examination* (1996), philosopher Paul Edwards cites the psychiatrist J. J. Preisinger Jr., himself an NDE survivor who underwent many of the experiences Moody describes. Preisinger wrote in the winter 1990–1991 issue of *Free Inquiry*:

> *When oxygen to the brain is reduced, as happens during the administration of anesthesia, drowning, and in altitude sickness, a great euphoria results, along with confusion and disorientation. People with hypoxia (decreased and insufficient oxygen) have the feeling that they have little or no concern as to the consequences of the deprivation . . . The instability of the temporal lobe of the brain, the major area that produces visual hallucinations, can well account for the light and tunnel visions. Shutting off oxygen intake to the brain when we are dying serves to eradicate the pain that most humans falsely believe to be intrinsically associated with death.*

In short, what Moody's interviewees appear to have experienced was not the fringe of the afterlife but merely the symptoms of the body shutting itself down, with the blood and oxygen supply to the brain decreasing as the system grinds to a halt. All the rest is the human mind trying to narrativize the events of the experience such that they tell a consistent story—or at least one that fits with our own prior understandings. We do not have to invoke the supernatural to explain the experiences Moody's interviewees describe.

Although the notion that, facing imminent death, you see the whole of your life flash before your eyes has come to have something of the status of an urban legend, interviews with people who've somehow survived events that by all rights should have seen them die— such as skydivers whose parachutes have failed to open, or climbers who've fallen from rock faces—show that many of them do indeed have this experience. The cause is not fully understood, but it is thought to have something to do with the way the brain stores far more memories than we normally have access to, and when the brain reckons that, well, this is *it*, it coughs up all those memories into the conscious mind, as if to ensure they're not wasted. People who undergo the more conventional form of NDEs—approaching death in a hospital bed or thrown into a coma by some trauma—seem not to encounter this process. Instead, according to Moody, they undergo a "life review" in the presence of a benevolent being. The parallel

⋙ A RETURN TO THE WOMB? ⋘

As we've noted repeatedly in this book, scientists speculating outside their own fields are as prone to error as the rest of us. The astronomer Carl Sagan suggested that all the imagery of the NDE—the long tunnel, the glowing light, the welcoming figures—is strongly reminiscent of the journey down the birth canal. Could the NDE be some form of memory of being born? In this model, the famous silver cord could represent the umbilical cord. It's a tempting idea. Alas, it founders on the fact that people who were born by Cesarian section are just as likely to experience standard-model NDEs as anyone else!

No, not an NDE or birth event, despite the tether: This Soviet stamp celebrates humanity's first spacewalk.

between this and the life-flashing-before-your-eyes experience seems very obvious, in that one seems to be merely a calmer, far more tranquil version of the other.

❧ HEAVENLY TOURISM

The derogatory term *heavenly tourism* (or *heaven tourism*) seems to have been coined within the Christian community to describe the many claims by people who have undergone NDEs that they have visited heaven and interacted with angels, dead people, Jesus, Satan, and even God. In a sense, many recorded NDEs could also be thought of as falling into this category, in that, as we've seen, very often people have reported having had a chat with God or a godlike figure during which they were required to reevaluate their lives.

It would be easy to assume that the current heavenly tourism fad began with Nebraska preacher Todd Burpo's *Heaven Is for Real: A Little Boy's Astounding Story of His Trip to Heaven and Back* (2010), a *New York Times* best seller that was made into a relatively successful movie in 2014. In fact, though, the fad can be traced back far further. Among its precursors, we find:

- ◆ *My Descent into Death* (2000) by Christian preacher Howard Storm, who had an NDE in Paris in 1985 while awaiting emergency surgery. Storm apparently had a stint in hell before being rescued and taken to heaven by Jesus Christ.
- ◆ *90 Minutes in Heaven* (2004) by Don Piper with Cecil Murphey, in which we're told that, after a car crash, Baptist preacher Piper spent some time in heaven meeting angels and long-dead relatives.
- ◆ *23 Minutes in Hell* (2006) by Christian preacher Bill Wiese, who apparently believes that in 1998 he had an afterlife experience in the "Other Place."

GEORGE RITCHIE'S PRETERNATURAL VOYAGE

Far earlier than any of these was a tale first told in a chapter of the book *To Live Again* (1957) by American inspirational writer Catherine Marshall and later in much more detail by the experient himself (although still retaining anonymity) in the December 1970 issue of the magazine *FATE*. Finally, the individual revealed himself to be the American psychiatrist George G. Ritchie; under his own name he wrote the books *Return from Tomorrow* (1978, with Elizabeth Sherrill) and *My Life after Dying* (1991). The latter book had what we might call a second lease on life in 1998 under the title *Ordered to Return*. The story he had to tell was truly an astonishing one.

On December 21, 1943, Ritchie died of pneumonia in an army hospital at Camp Barkley, Texas. Although the attending physician declared him dead, a ward attendant pleaded with the doctor to give Ritchie one last chance. To the amazement of everyone in attendance, an injection of adrenaline into Ritchie's heart did the trick, and eventually he made a complete recovery. It was during the nine minutes between Ritchie's death and the start of resuscitation that Ritchie conducted his tour of the afterlife.

Not realizing that the corpse on the bed was his, Ritchie decided to make for Richmond, Virginia, his intended destination before being taken ill. He flew in that direction at colossal speed, beginning to wonder how he could do this. Pausing midway at a town that he later discovered was Vicksburg, he discovered the ability to fly wasn't the only change—he was also insubstantial. When he tried to intercept a man who was going into an all-night café, it soon became apparent that the man could not see or hear him, and could not feel his touch. So Ritchie flew back to Camp Barkley, and this time, on seeing his own corpse, he realized the truth. Just then, Jesus Christ appeared, and Ritchie saw his whole past spooling before him as Christ asked, "What have you done with your life?"

The first stages of the posthumous journey on which Christ then took Ritchie involved visits to various sites where earthbound souls stalked the living. For example, they witnessed a female soul who had paid with her life for her nicotine addiction begging the workers on an assembly line to give her a cigarette, frustrated because none of them knew she was there. Next, the two travelers saw a boy's spirit pleading with a living girl for forgiveness, and perpetually thwarted because she was unaware of his attentions; this spirit was of someone who had committed suicide and who would forever be "chained to every consequence of his act"—all very *A Christmas Carol*.

Then Christ took Ritchie to other dimensions. In the first they saw a receiving station for those whose religious beliefs required them to sleep through the whole of their time in the afterlife until the Second Coming. The next stop was in hell, where wildly quarrelsome spirits were trying to fight to the death over trivial points or just yelling abuse at each other; some were even attempting to have sex with each other. Ritchie realized that these spirits had been so locked into their worldly desires during life that the obsessions continued after death, even though now pointless.

The spirits of hell couldn't see Christ and Ritchie,

John Leech's depiction of a ghostly apparition in Dickens's *A Christmas Carol* (1843). Marley came back from hell to warn Scrooge of what awaited him.

and the same was true of the spirits in the next dimension they went to. These spirits looked somewhat monastic and were working in a huge university-like environment on research projects and artistic activities. A highlight of the visit was a venture into a vast library that contained all the important books of the universe. Ritchie initially thought this dimension must be heaven, but then concluded that it was actually the destination for those souls who, in their mortal lives, grew beyond selfish desire.

The final destination was a city in remote outer space. This city was a blaze of light—the light of the dominant focus of life there: love. This was where people went who had achieved Christ-like status while they were alive. This was heaven, Ritchie knew, and he yearned to enter it. But instead he had to return to Earth and to his own mortal body.

Well, corroborate *that*. To his credit, Ritchie was aware of this problem: It was obviously going to be difficult even to find witnesses of his earlier brief journey to Vicksburg and back. But wait! He had seen that all-night café . . .

In the *FATE* article, Ritchie recounts how, almost a year after his NDE, he and a fellow student found themselves driving through Vicksburg—and suddenly Ritchie recognized the place as the city he'd visited when trying to reach Richmond during his initial astral journey. He guided his friend to the all-night café he'd seen while in his astral body. In *Return from Tomorrow*, the story is different: Now there were four friends in the car, and Ritchie did not tell any of them how excited he was to discover that this was the city of his vision, nor did he point out the all-night café. Which of the two versions is correct, if either? It hardly matters. The mere fact that there are two of them means they're valueless as corroboration.

But might there be anything in the other-dimensionly sojourns that could be related to real-world events? Well, it seemed to Ritchie that, yes, this could be so. About a decade after his astral adventure,

Dante and Beatrix hover to the right of St. Thomas Aquinas and Albert the Great above an assemblage of saintly scholars in this illustration from Dante's *Paradiso* (written before 1320). Perhaps Ritchie read Dante prior to his visits to the divine university and city of light?

according to the *FATE* interview, Ritchie was leafing through a copy of *Life* when he saw a photo of a control panel aboard the world's first nuclear submarine, the *USS Nautilus*. He knew he'd seen it before! And then the answer came to him: Some of those monklike scientists in the quasi-university he had visited with Christ were working on an instrument console exactly like this one. Not corroboration, but at least a striking concurrence.

In fact, according to Paul Edwards's extended analysis of Ritchie's "corroborations" in *Reincarnation: A Critical Examination* (1996), the relevant article in *Life* was about not *Nautilus* but its sister ship, *Seawolf*, but this is a minor discrepancy—anyone could make that mistake. A discrepancy that *isn't* so minor is that the article doesn't contain a photo of a control panel.[9] By 1978, however, Ritchie's memory had changed. Instead of coming across a photo of an instrument panel in *Life*, he'd found a schematic cutaway of a spherical vessel, and instead of having seen an instrument panel in

---

[9] It's feasible, of course, that Ritchie misremembered which magazine he had read.

The cover of the 1896 tract *The Roads to Heaven and Hell* indicates the two routes—
one filled with civil interaction and leading to a legless bearded man bathed in light,
and the other filled with fracas and leading to a pit of fire. While various people
claim to have visited heaven, returns from hell are far less common. Ritchie was
lucky enough to get a tour of both!

the astral dimension he'd seen a . . . But you're way ahead of me, aren't you?

Again, having two different versions of the same story means that neither of them corroborates or supports anything.

Did Ritchie's NDE transform his life? It would seem almost certainly so. But nowhere here is there any evidence that the afterlife exists.

MIRACLE OR MALARKEY?

Todd Burpo's short, best-selling book of 2010 tells how his three-year-old son, Colton, suffered a ruptured appendix, nearly died, and months after his recovery began to talk about his experiences in heaven. Among other things Todd explains that Jesus has a horse, and that its coat is of rainbow colors. Apparently Colton also met Jesus's cousin—presumably John the Baptist—and thought he was a nice guy. It's obviously very hard to evaluate material like this.

More details have emerged about the genesis of another book with a similar premise, published that same year. Alex Malarkey was just six in November 2004 when the car in which he was traveling, driven by his father, Kevin, was involved in a horrific crash. Kevin survived without major harm, but Alex was so severely injured that he remained in a coma for two months and is today a quadriplegic. On awakening from the coma, he claimed he'd seen his father hurled clear of the crash and caught by an angel.

According to *The Boy Who Came Back from Heaven: A True Story* (2010), published as a coauthorship between Kevin and Alex, Alex died and sojourned in the afterlife, being escorted to heaven by angels and meeting people there, including "white angels with fantastic wings, green demons with long fingernails and hair made of fire"—not to mention Jesus Christ and Satan, whose "mouth is funny looking, with only a few moldy teeth." Apparently there's a "hole in outer heaven" that Satan can pop through to travel to and from hell.

Though Alex also encountered God, he was not permitted to view the deity's face.

Even after emerging from his coma, Alex made occasional repeat trips to Paradise. While the book did not match the success of Burpo's *Heaven Is for Real*, it still reportedly sold well until the publisher stopped selling it in January 2015, when Alex himself released a public statement declaring the account to be in essence a fabrication.

Alex Malarkey's account of Satan—which also described a three-headed devil with hair like fire—bears a vague resemblance to but lacks the menace of Dante's version, depicted in this fourteenth-century illumination. The more mouths you have, the more people you can chomp at a time.

"PROOF" OF HEAVEN

An even more recent example of heavenly tourism is the book *Proof of Heaven* (2012) by neurosurgeon Eben Alexander. According to Alexander's account, he woke up one morning with a severe headache, was rushed to the hospital, was diagnosed with a rare bacterial meningitis ("*E. coli* bacteria had penetrated my cerebrospinal fluid and were eating my brain," he told *Newsweek* in October 2012), and fell into a coma that lasted seven days.

Then, just as the medics were considering whether to discontinue life support, he regained consciousness and, after convalescence, started to recount the astonishing adventures his mind had undergone while he was "brain dead." Allegedly, while visiting heaven and witnessing all the marvels there, he saw transcendental winged beings, higher life-forms that he identified as angels. He heard beautiful song that seemed to fill the universe. There were butterflies everywhere, and for at least part of the time he was riding on a butterfly's wing alongside a lovely young woman. This young woman stayed with him throughout his travels in the afterlife. Later, long after he had returned to his mundane existence, Alexander—who had been adopted in infancy—was able to identify her as a sister whom he'd never known he had, a sister who died in childhood.

Alexander casts himself as a sober scientist who was well aware of the research done on NDEs and had always accepted the conventional explanations for them. But his case, he insisted, was different: The experience could hardly have been born from the brain firing off in all directions because his brain at the time had been dead, gobbled up by all those bacteria.

But if consciousness—mind—can exist without the brain, then much of modern science needs to be rewritten. We need to go back to the drawing board and create a new picture of reality—one that incorporates different ethereal dimensions, an entirely new understanding of the role of the human mind in the functioning of the universe, and much more besides. Or we can conclude that Alexander got some of his science wrong, which is the view taken by neuroscientist Sam Harris after he saw the *Newsweek* piece. In a lengthy post on his blog,[10] Harris explained that, in principle, he was prepared to believe that consciousness could exist separately from a functioning brain, simply because so far we have little understanding of how consciousness happens. At the same time, though, he found severe problems with Alexander's science. Apart from the facts that (a) if your brain is dead then so are you, and (b) destroyed cortical tissue doesn't regenerate, Alexander seemed to believe that CT scans and "neurological examinations" were proof that his cerebral cortex had been "stunned to complete inactivity." The trouble is that CT scans and neurological examinations are incapable of showing whether or not brain activity continues—for that, you need something like an EEG.

---

[10] "This Must Be Heaven," October 12, 2012. http://www.samharris.org/blog/item/this-must-be-heaven.

Did Alexander ever read Richard Doyle's classic *In Fairyland* (1870)? This illustration of a boy and girl fairy flying among a flock of butterflies suggests as much.

The most likely scenario is that Alexander could simply have been dreaming during his coma, or even experiencing visions during the relatively short time when he was coming out of it. And yet, he claims, the experiences were far too rich and vivid for this to have been a possibility. He also rejects suggestions that this vividness might have been due to psychedelics like DMT (N, N-Dimethyltryptamine), which is found in the brain and is thought to have a role as a neurotransmitter, or ketamine, which is used in anesthesia and can induce visionary trances. DMT is found not just in the human brain but widely elsewhere in nature—for example, in various plant species common in Amazonian Peru. The people there and elsewhere in South America have known about the psychedelic properties of these plants since time immemorial and use their leaves as the active ingredient in the psychedelic tea known as ayahuasca (and under various other names). Ayahuasca is employed for ritual purposes because of its ability to conjure up mystical visions; the tea's purgative properties are also regarded as spiritually valuable. Among those institutions making wide use of ayahuasca is the Brazilian church called Uniao do Vegetal (Union of the Vegetable).

As a psychedelic drug, DMT has been studied since the mid-1950s, when the Hungarian Stephen Szára investigated its properties using the services of experimental volunteers. In the United States it played a small part in the prevalent drugs culture of the 1970s, earning itself the nickname "businessman's trip" because its effects were so short-lived: You could start tripping at the beginning of your lunch hour and be perfectly competent to resume work punctually in the afternoon!

Among the other effects Szára reported was that his volunteers not infrequently described how, while tripping, they seemed to encounter strange, otherworldly entities. This has been widely reported elsewhere since. According to psychiatrist Rick Strassman, in *DMT: The Spirit Molecule* (2001), about half of his experimental DMT subjects went into a deeper phase of hallucination, which they experienced as essentially an independent, alternate existence—something that seems not unlike reports of the astral plane. And it was in that parallel existence that they encountered what seemed to be independently intelligent "guides." These guides could take all sorts of forms, from stick figures to animals to bizarre aliens—even to intelligent plants! Later, the psychonaut Terence McKenna dubbed these DMT-induced beings "machine elves." An additional characteristic of these experiences is that many of the subjects remained convinced that the creatures they had encountered while tripping genuinely existed, albeit on some different plane.

The people who underwent such experiences were using relatively high doses of DMT; the DMT present in the human brain is at lower levels—otherwise we'd all be in a sorry state (though possibly a happy one). Even so, it's clear that there are some interesting

⋙ IT'S NOTHING TO WORRY ABOUT ⋘

A major disadvantage of dying is that people tend to get a bit frightened as the final curtain approaches. But fear not! Help is at hand, thanks to Robert A. Monroe, an ex-adman who became an OOBE expert, founded the Monroe Institute of Applied Science, and wrote the books *Journeys Out of the Body* (1971), *Far Journeys* (1985), and *Ultimate Journey* (1994). In 1975, he invented an audiovisual device that made use of a process he'd devised called Hemi-Sync (short for Hemispheric Synchronization). This involves using recordings of sound pulses in order to stimulate (supposedly) the separation of the soul from the body. The purpose of Hemi-Sync is to launch the user into altered states of consciousness. Most excitingly, the device can be used to give dying patients sneak peeks at the afterlife and thereby ease any worries they might have about the prospect of spending a long period there. So successful was the device that Monroe reported most of those who tried it decided to stay in the afterlife, rather than return to their bodies to live out the last few weeks or months on Earth.

This illustration from *The Pied Piper of Hamelin* (1888) parallels the seduction of Charles Palmer's bucolic afterlife.

One of the odder Spiritualist publications, Charles Palmer's *Spiritual Truth for the Young* (1950),[1] has a similar concern. It explains to young children what death is like in case the prospect is making them wake up screaming in the middle of the night. Seemingly, once you're dead, you can give yourself ice cream or a bar of chocolate just by willing it: Now *there's* a thought to soothe many a youthful nightmare. And there's also the thought that things are going to look up a bit on the medical front, too, because:

> . . . the etheric body is always perfect. If the physical body is not quite as it should be . . . the etheric body has none of these troubles. So you can see that, when a cripple passes over, he finds he can walk and run just like other people; and a blind man finds he can see perfectly. Even one of those poor unfortunate people whose minds are out of order, on passing over, is able to think, and speak, and act, as sensibly as anyone else, because he is no longer hampered by an imperfect brain. Think what it means to a poor cripple, or blind man, to know that when he passes over, he will regain, in full perfection, his lost faculties!

This sounds, of course, like a recipe for suicide the next time the youngster has a bad cold.

On the downside, there's school in the afterlife; on the upside, the lessons are fun; on the downside, they're only fun if you think ceaseless religious instruction is fun.

When school's let out, however, there are all sorts of permissible ways you can communicate from the afterlife with your old friends and, presumably, scare their little pants off in the attempt.

---

[1] Discovered by Alfred Armstrong and described by him on his excellent *Odd Books* site: http://oddbooks.co.uk.

parallels between the DMT experience and many accounts of NDEs and visits to heaven. At this stage, to say more than that might be speculating too far.

Harris concluded his article thus:

> [T]his issue of Newsweek *is best viewed as an archaeological artifact that is certain to embarrass us in the eyes of future generations [and] reveals the abasement and desperation of our journalism, the intellectual bankruptcy and resultant tenacity of faith-based religion, and our ubiquitous confusion about the nature of scientific authority . . . I hope our descendants understand that at least some of us were blushing.*

Of course, the science-based criticisms of Alexander's book by Harris and others—including Oliver Sacks—did nothing to curb its sales.

There are two concerns in this case. First, did Eben Alexander undergo an NDE? There seems no compelling reason to deny that he did, even if the details of how it came about—during the coma or on emergence from it—are very much open to question. Second, did he actually spend time in the afterlife? Probably not. There are various plausible scientific explanations, as briefly noted above, for what he experienced. This is not to belittle the importance of the experience to him, which seems to have been great.

The psychedelic drug DMT is the mainstay of the Peruvian ritual brew known as ayahuasca. This may look like an innocent stir fry in the making, but . . .

By all accounts, NDEs can be profoundly affecting for those who have them. But that fact in itself does not represent evidence for the existence of an afterlife. In the words of the English classical scholar and poet A. E. Housman, "The house of delusions is cheap to build but draughty to live in."

The first time I walked my young daughter to school, we had this conversation:

> *Jane: "I just saw God."*
> *Me (after a long search for words): "Where?"*
> *Jane: "Too late now. We've gone past Him."*

Maybe I should have written a book about it.

4

ONCE IS NOT ENOUGH

PAST
AND
FUTURE
LIVES

"It is not more surprising to be born twice than [to be born] once."

—Voltaire, *The Princess of Babylon* (1768)

"Maybe remembering [past lives] is a defect. Maybe we're supposed to forget, but sometimes that system malfunctions, and we don't forget completely."

—Ian Stevenson, cited in Tom Shroder's *Old Souls* (1999)

I N HIS EXCELLENT BOOK *PARANORMALITY* (2010), Richard Wiseman recounts an investigation he undertook in Hampton Court Palace. The palace has long been thought to be haunted by various spirits—most notably that of Catherine Howard, the fifth wife of Henry VIII. Although Henry described her around the time of his marriage as his "rose without a thorn," within a couple of years he was beheading her on the grounds of her adultery. In 2001 Wiseman was called to the palace because of a recent increase in Catherine's spectral activities. He and his team enlisted the help of the public—one of whom latched onto them and wouldn't go away, declaring herself to be a reincarnation of the beheaded queen and therefore able to offer unique specialist assistance to the investigation.

The team had rigged up heat sensors and a thermal-imaging device in the corridor where Catherine's ghost was most often reported. Reviewing the heat-sensor data one morning, the team discovered that there had been an upsurge in local temperature a few hours earlier, at around 6 a.m. The thermal imager, too, recorded a figure entering the corridor—and Catherine's "reincarnation" promptly identified it as a courtier whom she'd known in her previous life. Unfortunately for her claim, the "courtier" could then be seen pulling out a vacuum cleaner and setting to work with it . . .

An engraved portrait of Henry VIII's fifth wife, Catherine Howard, above an angel with an extinguished torch seated next to her severed head. Does her ghost haunt Hampton Court? Er, probably not.

The idea of reincarnation requires us to accept the notion that there is life after death. To most Americans, other than those in New Age circles, reincarnation seems a highly unlikely possibility, and so it's often regarded as some sort of outlier. Yet this is a very "Western" point of view. While Muslims, Jews, and Christians tend to reject reincarnation, Hindus and Buddhists by and large accept it. How do you think the Dalai Lama is selected? And for centuries Hindus and Buddhists have far outnumbered adherents to the Abrahamic religions. This is a point worth remembering whenever Christians or others try to use the argument that there must be an afterlife because most people throughout history have believed so. If that's the case, then by the same reasoning, all of us should accept reincarnation as a fact.

Alternative terms used to describe reincarnation include *metempsychosis* and *transmigration of souls*. Related concepts include the *astral body* and the *astral plane*. Touched upon briefly in the previous chapter, the astral body is a supposed second body that we all have, usually assumed to be identical to our physical one but insubstantial (and almost but not quite always invisible). This body is capable of travel in the astral plane, a level of existence usually assumed to be "above" this one, that's accessible via death, trance states, and the like. Needless to say, since by definition the astral plane and associated phenomena cannot be detected by any scientific instruments, their scientific status is—to be charitable—dodgy. Some photographs do exist purporting to show astral bodies, usually in ectoplasmic splendor, but every time they've been subjected to skeptical analysis, they've been found to be fakes.

Not everyone agrees that the astral body must resemble the physical one. In *Life after Life* (1975), Raymond Moody (page 143) is less dogmatic, ascribing to it:

> . . . *a form or shape (sometimes a globular or amorphous cloud, but also sometimes essentially the same shape as the physical*

body) and even parts (projections or surfaces analogous to arms,
legs, a head, etc.). Even when its shape is reported as being
generally roundish in configuration, it is often said to have
ends, a definite top and bottom, and even the parts just
mentioned . . .

Most of us don't retain our memories from one life to the next—
which is just as one might expect, because memory is a function of
the physical brain. Indeed, since the astral body's brain must be as
ethereal as the astral body itself, it's difficult to see how anyone could
retain memories through the interval between incarnations.

If reincarnation were a reality, then we'd expect the "reincarnation
experience" to be the same for everybody, no matter where in the
world they lived. In fact, details vary quite radically, depending on the
religious and cultural beliefs of the society in which the claimant lives.
For example, in most cultures it is assumed that the individual will
experience a gap of weeks, months, years, or longer between the end
of one incarnation and the start of the next, and this is exactly what
is reported from those societies. Among the Jains, however, it's
believed the new incarnation follows immediately after the end of the
old, and, once again, this is exactly what's reported. Again, some
cultures record that the new incarnation will occur relatively close,
geographically, to the old one, but in others the separation can be
hundreds of miles. Among the Alaskan Tlingits, reincarnation occurs
only within the same bloodline.

In his book *The After Death Experience* (1987), Ian Wilson
discusses two UK hypnotists whose specialty was regressing subjects
into their past lives. One of them, Joe Keeton, believed that concep-
tion of the new person occurred at the same instant as the physical
death of the old, and he informed his subjects accordingly before
putting them under. Sure enough, the subjects reported exactly this.
The hypnotist Arnall Bloxham, by contrast, issued no such advance

notice, and his subjects "exhibited 'past lives' that hopped cheerfully from the Stone Age to ancient Egypt, to ancient China, to Inca Peru, to medieval Bavaria, and ultimately to South Wales, with often centuries-long intervals between." Again, preconceptions appear to dictate what's reported as reality.

While many modern researchers appear all too ready to accept cases of supposed reincarnation at face value, those doughty early investigators of the SPR were significantly less tractable. In a case dating from 1906, for example, they examined a woman called "Miss C" who, under hypnosis, could "recall" having previously been one Blanche Poynings, a close friend of Maud, countess of Salisbury, during the reign (1377–1399) of Richard II of England. As far as the details of the countess's life could be checked, it seemed Miss C's "recollections" were remarkably accurate—and she swore that she'd never read a history book or historical novel that might have given her the information.

One day, while she was engaged in conversation with an SPR investigator, the subject turned to the use of the planchette, and Miss C declared herself ready to give it a try. A happy dialogue ensued between Miss C and Blanche Poynings, in the course of which Blanche was asked for a way of checking on her story. She replied, by means of the planchette, "Ask E. Holt." Some detective work revealed that the author Emily Holt had written a children's novel, *Countess Maud* (1892), about the countess of Salisbury, and every detail of the countess's life that Miss C had recounted under hypnosis proved to have come from this novel. Although Miss C had insisted with complete honesty that she couldn't remember having read any books that might have been responsible for her knowledge of the period, it turned out that she had read Holt's novel years ago in childhood and had completely forgotten having done so.

The sequence of animals depicted on this Tlingit totem pole in Wrangell, Alaska, hints at the tribe's reincarnation beliefs.

⇒ IAN STEVENSON ⇐

In the modern era, the doyen of reincarnation investigators was Ian Stevenson. Born (this time around, at least!) in Montreal in 1918, he graduated from the McGill University School of Medicine in 1943 and eventually made psychosomatic medicine and psychiatry his specialties. Later he moved to the University of Virginia, where he studied (and sampled) LSD and mescaline, the former of which gave him three days of "perfect serenity"—an experience that some have suggested may have contributed to his burgeoning interest in reincarnation. His first publication in the field was "The Evidence for Survival from Claimed Memories of Former Incarnations" (1958) in the ASPR's *Journal*. In 1961, he started making field trips to areas of the world where the belief in

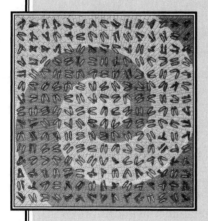

reincarnation is especially profound, producing a string of books reporting cases of supposed reincarnation, starting with *Twenty Cases Suggestive of Reincarnation* (1966; revised 1974). Also of relevance are his studies of xenoglossy (see page 195) and biological indicators—for example, birthmarks supposedly carried over from one incarnation to the next.

Although Stevenson was widely regarded as an objective investigator, questions were

✴ CELEBRATED REINCARNATES

General George S. Patton was convinced that one of his previous incarnations was as the Punic military commander Hannibal. This is a notion enthusiastically endorsed by the website Return of the raised as to his impartiality. Rather than faking his results, it seemed that he might be letting his eagerness to believe in reincarnation persuade him to accept as genuine cases that, to other observers, appeared more dubious. For example, in *The After Death Experience*, Ian Wilson cites the case of a little Delhi boy, the son of a poor shopkeeper, who was claimed to be the reincarnation of an extremely rich local philanthropist who had died some years earlier. However, the child did so poorly at his interview with the philanthropist's survivors that they rejected any possibility of the claim. Later, a local doctor explained to Stevenson that the boy had been coached—obviously inadequately!—by a neighboring resident in hopes of passing him off as the rich man's reincarnation. Anyone else looking at the facts in the case would have dismissed it as value-less, but Stevenson decided to discount the testimony of the doctor and cast doubt on the veracity of the philanthropist's family!

Wilson also points out that, in virtually all the cases discussed by Stevenson, the claimants are near the bottom of the social and financial heap, while the families to which they claim their former incarnations belonged are substantially better off. Of course, one could conjecture that children reborn into wealthy surroundings might choose to keep quiet about their poverty-ridden former families, but wouldn't the lucky children want to share their good fortune with the people they could remember loving in their previous lives? Isn't it more likely that, in communities where the belief in reincarnation is strong, poor people see the claim that a child "really" belongs to a rich family as a means of material and social advancement?

Stargods (http://stargods.org), which even offers comparative images so you can see the resemblance for yourself. These keen investigative journalists, oddly, spell the Carthaginian general's name "Hannible."

It's been suggested that Patton's faith in reincarnation was a mechanism he used to control his fear while in the heat of battle. He certainly believed his many incarnations were almost all as a warrior. Another of his supposed incarnations was as a Roman legionnaire. In his long poem, "Through a Glass Darkly" (1922), he wrote:

> *Perhaps I stabbed our Savior*
> *In His sacred helpless side,*
> *Yet I've called His name in blessing*
> *When in after times I died.*

There's a tale that, while in France during World War I, he was able to show others around Langres, pointing out to his companions such Roman-era highlights as the forum and the site where Caesar had pitched his tent—even though, in this life, he had never been there before. Patton himself seems far more tentative on the subject in a November 1917 letter he wrote from France to his mother:

> *I wonder if I could have been here before. As I drive up the*
> *Roman road the Theater seems familiar—perhaps I headed a*
> *legion up that same white road . . . I passed a château in ruins*
> *which I possibly helped escalade in the middle ages. There is no*
> *proof nor yet any denial. We were, we are, and we will be.*

The British mystic Paul Brunton (born Raphael Hurst) had previous incarnations, he claimed, on worlds other than Earth. One was on the planet Venus—this before mere earthly science revealed that the surface of Venus was hotter than an oven—and another was on a planet of the star Sirius. Among his revelations about the first of these worlds[1] is: "There are no cars on Venus." As for the latter:

---

[1] As cited by Colin Wilson in *Rogue Messiahs* (2000).

Hannibal—or is it George Patton?—crosses the Rhône in this 1878 impression by French artist Henri Motte. Below, George Patton—or is it Hannibal?—stands in front of a more modern military transport vehicle in France, 1918.

The beings who people it are infinitely superior in every way
to the creatures who people earth. In intelligence, in character,
in creative power and in spirituality we are slugs crawling at
their feet. The Sirians possess powers and faculties which we
shall have to wait a few ages yet to acquire.

Toward the end of his life, the science fiction writer Philip K. Dick (he died in 1982) started having vivid, spontaneous recollections of previous lives in ancient Rome and elsewhere. These seem to have been hallucinations linked to his drug and alcohol abuse, stress, sleep deprivation, and even the use of sodium pentothal during dental procedures.

Among those who claimed during the mid-1970s that they could recall their previous existences were the Virginia preacher's wife Dolores Jay, who remembered being Gretchen Gottlieb, murdered a century before in Germany, and Gavin Arthur (Chester Alan Arthur III), a grandson of US president Chester Arthur, who recalled living in the thirteenth century. The avant-garde silent movie *Borderline* (1930) offers us an opportunity to see both Gavin Arthur and his first wife, Charlotte, playing opposite American singer/actor Paul Robeson. A bisexual and a practicing astrologer, Arthur published the successful book *The Circle of Sex* (1962), in which he claimed there were no fewer than twelve different sexual orientations—one for each sign of the zodiac.

In his book *A Door to Eternity* (1979), the Australian novelist G. M. Glaskin—whom we earlier encountered when talking of the Christos Experience (page 137)—describes a number of his non-Christos dreams, two of which he felt might have been, at least in part, "reincarnatory." In one dream he was a medieval monk who fled when his fellow monks for some reason began killing each other; in the other he was a sailor arriving onboard a very ancient steamship and being given the runaround by an officious junior officer.

Glaskin offered—and rejected—an entertaining rationale for reincarnation. The Cathars and other heretical sects had the notion that, while the conventional wisdom has it that hell is the place bad people go when they die, in actuality we're already *in* hell. Bearing in mind what the Catholic Church did to the Cathars, this couldn't have seemed too far-fetched. But the Church also put forth the idea of purgatory, where souls go to be cleansed of their lesser sins. Putting these two notions together, you come up with what is surely a far more attractive model: We live in purgatory! Clearly, most souls require several passes through purgatorial misery before they're ready to go on to a better place.

The Northamptonshire chiropodist Jenny Cockell represents one of the more celebrated recent cases of claimed reincarnation— "celebrated" because she's been able to track down the children (now older than she is), whom she supposedly bore in her previous incarnation as Irish housewife Mary Sutton. At least one of them—Sonny, the eldest child of Mary Sutton—accepted her as his long-dead mother. Although Cockell failed to recall critical details of Sutton's life, including Mary's maiden name and the names of her husband and children, her story was the subject of the book–turned–TV movie called *Yesterday's Children* (2000), starring Jane Seymour.

SHIRLEY MACLAINE

Almost certainly the world's most famous reincarnate in recent decades has been the actress Shirley MacLaine, who has described her discoveries of this and other aspects of the paranormal in books like *Out on a Limb* (1983), which was made into a five-part TV miniseries in 1987, bringing her a Golden Globe nomination. In one of her previous lives, she claimed she became the court jester to Louis XV, but her impertinence overstepped the mark and the infuriated king cut off her head. It was an incident she remembered well, once the window had been opened to her long past. "I watched my head rolling on the floor," she

Shirley MacLaine, perhaps the most renowned of all the celebrity reincarnates, September 1987.

later reportedly recalled. "It landed face up, and a big tear came out of one eye." It's hard to tell which is the greater mystery here: (a) why history fails to mention the rather plodding Louis XV personally beheading anybody, or (b) from where MacLaine was watching her own head.

Another former life saw her as a male native of Atlantis. And not just any male: She was the brother of Ramtha, a warrior chieftain who currently resides in heaven and "channels himself" through the Washington State mystic JZ[2] Knight, founder of Ramtha's School of Enlightenment. Having been raised and educated on a prehistoric continent, Ramtha is obviously ideally placed to impart information about scientific subjects that he understands far better than actual scientists do, like archaeology, cosmology, neuroscience, and quantum physics. As I noted in my book *Discarded Science* (2006):

> *To be sure, modern scientists have difficulty finding any evidence for the various scientific insights that Ramtha offers, but this is their fault, not his; he promotes a non-evidential form of scientific research that can probably best be summed up as: "If it feels good, believe it." There is no objective reality, only subjective realities, so objectivity—such as orthodox scientific research—is necessarily a meaningless exercise.*

---

[2] This stands for "Judy Zebra," although she operates as JZ.

According to her own account, MacLaine was led into paranormal waters by a guru called "David," who she later revealed was a composite of four people who guided her. All have, in turn, been guided by their relationships with aliens who came here from the Pleiades. Since the stars of the Pleiades cluster are estimated to be only some 150 million years old, there doesn't

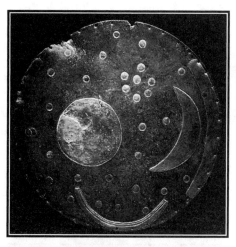

Aliens from the Pleiades star cluster are an important part of MacLaine's worldview. The Nebra sky disk showing the Pleiades dates from about 1600 BCE.

seem to have been very much time for intelligent life to have evolved on any planets there. Perhaps the Pleiades were just a port of call en route from the aliens' real home?

One of the astonishing paranormal events that MacLaine recounts in her book *Dancing in the Light* (1986) is an astonishing dematerialization that took place in broad daylight when she was in a shop in Beverly Hills. Apparently, she put her purse down, turned away for only a moment, and when she turned back again *the purse had disappeared!*

BORN-AGAIN WRITERS

When Louisa May Alcott, author of *Little Women* (1868) and other novels, was once asked why she had never married, she replied, "I have fallen in love with so many pretty girls and never once the least bit with any man." Elsewhere she explained her hypothesis as to why this might be so: In a previous life, "I must have been masculine, because my love is all for girls." Whether or not she was serious is hard to determine, yet as an explanation for homosexuality, this has attracted the attention of a number of reincarnation believers.

Various writers have based their careers, or at least individual works, on their belief that they can remember their previous existences. Perhaps the highest-profile writer reincarnate has been Joan Grant. It began in 1937 with her first "autobiography of the soul," called *Winged Pharaoh*. Like its successors, this book was, she claimed, merely the reporting of what she saw of her past lives when shifting to a different "level." Some of her rather matter-of-fact accounts of what it's *like* to

⇒ THE DALAI LAMA ⇐

Seventh Dalai Lama:
Kelzang Gyatso
(1708–1757).

There's another reincarnated soul who is internationally even more famous than Shirley MacLaine. Until the Chinese takeover in 1951, Tibet had since the late fourteenth century just a single ruler—the Dalai Lama—in different incarnations. The current Dalai Lama, born in 1935 and recognized in 1937, is the fourteenth, and he has stated in newspaper interviews that he may be the last. Comically, the Chinese government has declared that it's not up to him to decide whether or not he should reincarnate; the decision is theirs, as is the power to decide the identity of the new incarnation.

be someone who can supposedly remember past lives are oddly convincing—more so, to be honest, than the novels themselves.

Like MacLaine, Grant learned what it was like to have her head cut off. In *Time Out of Mind* (1956), she described some of her earliest "reincarnatory dreams," dating from when she was about seven years old. In these dreams she was a French girl, usually about ten or eleven years old but sometimes in her teens. The girl lived in "a house with a steep grey roof and faded green shutters in which most of the rooms were shut up." At one point, in real life, the young Joan Grant was peeved because her parents declined to take her with them to Paris. Surely the French girl of her dreams had been there! Indeed, yes, as Grant soon found out. At the age of nineteen, the dream-girl was taken to Paris and thrown alone into a dark dungeon, where she tamed a rat for company. The only time she ever saw sunlight again was during the short ride aboard a tumbrel to Madame Guillotine . . .

> It might have been a terrifying dream, but oddly enough it was rather comforting; because now I knew that beheading does not hurt at all. There was only a loud thud and a feeling of falling head over heels. The next minute [the French girl] was jumping over a stream to join two men who were waiting for her on the other side of it . . . two men whom she very much loved.

Grant contended that in ancient times everyone knew about reincarnation, and so death was feared less—even if you were dying painfully, you knew that, as soon as the pain was over, you'd be on the brink of a new life.[3] More recently, however, our knowledge of the many-life cycle has been lost, although you may still retain vague

---

[3] A similar notion was put forward by the reincarnationist Sir Alexander Cannon as an argument against the death penalty: What punishment can it possibly be if you're merely releasing the guilty party to another life?

⟹ THE SUPRA-PHYSICAL ⟸

According to Joan Grant's *Many Lifetimes* (1967), written with her husband Denys Kelsey, four distinct phases characterize one's reincarnations, even though an individual may incarnate many times within each phase. In stage 1 you—or, rather, your "supra-physical"—have the capacity to organize only a single molecule. However, you then start slowly working your way up through the mineral kingdom until you get to the plant kingdom (stage 2); next comes the animal kingdom (stage 3), and finally you become a member of *Homo sapiens* (stage 4). Your early human incarnations involve your entire personality, but in later lifetimes become so developed that there's no longer sufficient space (as it were) in the single human consciousness for the whole of your personality—hence the notion of the supra-physical, a part of which is manifested through you yourself and the rest of which seems to hang around in a kind of ethereal way, exerting a limited influence on you and your actions.

The trick of remembering prior lives seems to be to get in touch with your supra-physical. Any pregnant woman knows there are times when just the thought of even her favorite foods can bring on major nausea. This is not a simple matter of the body knowing better than its owner what foods it needs; rather, the unborn child's supra-physical is trying to communicate to the mother that it would rather have coal than cereal for breakfast.

memories of your previous deaths. An alcoholic in this lifetime, for example, perhaps previously died of thirst in the desert, or waited in agony for an operation in one of those primitive military field hospitals where the sole anesthetic was alcohol. Lying on your bed of pain, you watched the booze making the rounds and yearned for it with all your being. No wonder, then, that in this life your supra-physical can't get enough of the stuff!

Joan Grant's vision of the progression of reincarnations that the individual undergoes seemed to reflect some of the evolutionary debates
going on at the time. Here, discredited biologist Ernst Haeckel's 1874
illustration of the human pedigree as a Great Chain of Being—with
lemurs evolving directly from kangaroos—is similarly inaccurate.

Since these prebirth traumas may be having a deleterious effect
on your current existence, Grant and Kelsey, who was a psychiatrist,
came to the conclusion that helping alcoholics and other mentally ill
people recall details of their past lives might open the way to some
otherwise impossible psychiatric cures. (We'll come back to this
notion later.) In *Many Lifetimes* they give some examples of their
successes. It is perfectly possible to take their word for it that the

therapy was successful without accepting for one moment that the patients were recalling past incarnations. For the patient, the experience might well have been like some form of psychodrama.

In later years Grant relied almost entirely on what she called "tuning-in" to recapture her past memories. The former incarnation whose lifetime was the basis for Grant's first "biographical novel"— Sekeeta, who became a female pharaoh, lived in Egypt during the First Dynasty—had dreams of *her* previous lifetimes. In *Winged Pharaoh*, Sekeeta records many of these dreams; we discover inter alia that she dream-traveled to Athlanta (i.e., Atlantis) at a period five thousand years before her own era.

Another one-time Atlantean was the popular American novelist Taylor Caldwell, who was likewise convinced that she had had many previous incarnations. Her novel *The Romance of Atlantis*, revised by

During the 1930s, the Hungarian architect Géza Maróti produced a six hundred–page cultural history of Atlantis; for some reason, it remains unpublished, but here is an illustration from it.

Jess Stearn for publication in 1975, was written when she was only twelve. Her proud father sent it to her grandfather, who was a book editor. As Stearn tells it, the grandfather,

> *promptly horrified, suggested the manuscript be destroyed immediately. He did not feel that any child could have produced so unusually mature a work, intellectually and philosophically. The only alternative that suggested itself was that she had borrowed freely from elsewhere. In a way, he was right. She had borrowed from the past, not knowing herself how she was dredging up that past.*[4]

In her postscript to the finally published version of the novel, Caldwell—by then in her seventies—tells how she has begun to dream again of Atlantis and her life there as the Empress Salustra. She adds that she was surprised the dreams started again because she had more or less forgotten that Stearn was readying the manuscript for publication. The dreams were of new scenes—that is, scenes other than those in the novel. She recounts three of them.

In her first dream, she was walking around her Atlantean estate, where various animals were part of the décor. She mentions "the pit in which squirmed captured reptiles of a breed unknown to this present world, and animals also unknown to us moderns"—a curious remark in that most of us, presented with an assortment of strange-looking beasts, would assume they were extant animals we didn't recognize, not necessarily extinct ones. (Besides, Salustra was alive just twenty thousand years ago—a mere tick of evolutionary

---

[4] Another child prodigy, the UK novelist Jane Gaskell, wrote several Atlantean novels during her teens that bear up well to rereading—better, indeed, than some of her adult work. So there's no real justification for either Stearn's or Grandpa's assumptions. A reading of *The Romance of Atlantis* suggests, however, another reason why Grandpa might have recommended the flames.

time.) In Caldwell's second dream, set seemingly just before the erup-
tion that brought about Atlantis's demise, Salustra was being taunted
by Signar, the man she loved but who loved her sister. By the time of
the third dream, Atlantis—and presumably the sister—had foun-
dered, and Salustra was now united in love with Signar as they and a
plucky band of survivors did their best to create new lives elsewhere.

EDGAR CAYCE

Edgar Cayce, the renowned Sleeping Prophet, could recall not only
his own lifetimes in Atlantis and Lemuria but also those of other
people. Moreover, he confidently predicted that these two lost conti-
nents would soon rise once more from beneath the waves. In the
1960s, advances in geology made it plain that, if ever there had been
a continent in the middle of the Atlantic, it must have been more
than two or three hundred million years ago—long before not only
the emergence of humankind but even the appearance of the first
primates. Even so, in his 1968 book *Edgar Cayce on Atlantis*, the
Sleeping Prophet's son, Edgar Evans Cayce, was continuing to press
the case, and was in particular pointing out all the cultural similarities
that can be found on both sides of the Atlantic. One of these, he
gasped excitedly, was that the people of both the Old and New
Worlds share a 365-day calendar!

Cayce *père* was, we're told, convinced of the reality of reincarnation
by an experience he had as a young man.[5] He dreamed of being one of
a band of white settlers fleeing on a raft down the Ohio River, where
he and his companions were eventually massacred by Native
Americans—which might well have been a kinder fate than death by
starvation, which was imminent. Some time later, Cayce was in a
barbershop when a small boy climbed onto his knee. The boy's father

---

[5] The account of this event presented here is based on that in Jess Stearn's *The
Sleeping Prophet* (1967).

said something about not pestering strangers, to which the child replied that Cayce was no stranger: "We were hungry together at the river."

If we had any reason to believe this story was true, it might serve as evidence in favor of reincarnation. Unfortunately, the level of mythopoeia surrounding Cayce is so extreme that it's impossible to take any story about him at face value. The same goes for his supposed recollections of lives in ancient Egypt as well as more recent historical periods.

Initially, according to Cayce's "memories," the human race was not physical but ethereal. Then humanity began to take on physical form and, some 98,000 years "before the entry of Ram into India" (whenever that was), humans invented sex. Cayce's explanation of which of these two developments occurred first is difficult to understand: Maybe the spiritual creatures thought it would be fun to have bodies in order to enjoy sex more, or maybe they became corporeal and—hey, look what I got! Whichever occurred first, the incarnating thought-forms took on the color of the stuff of which they were made—starting with the red rocks of Atlantis. According to Cayce, as far as I can understand, that's where the supposed "redness" of Native Americans came from.

Atlantis was destroyed in not one but three stages, the last of which took place around 10,000 BCE (this was the disappearance into the sea to which Plato referred in his *Timaeus* and *Critias* dialogues). Dating the three catastrophes is difficult. As Cayce *fils* pointed out, "The time of these events is so remote (over 10,000 years ago) that few records remain. Those that do are either undiscovered or unrecognized." Or, of course, nonexistent.

In his sleep, Cayce observed the details of many other people's previous existences in Atlantis. Such previous lives were believed to have profound effects on people's modern existences. Take the case of Robert Dunbar, who, according to Cayce, had previously lived in Atlantis, Egypt, India, and Germany—each time being involved in

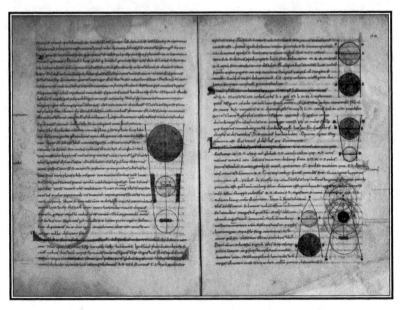

Medieval manuscript of Calcidius's Latin translation of Plato's *Timaeus*.

scientific or engineering work. In India, for example, he learned how to make explosives for use in warfare. In his current existence, Dunbar was still a small child when all this past history began to come out, and Cayce instructed the boy's parents to ensure that in adulthood their son's potentially dangerous abilities would be put to good rather than destructive use. Indeed, after working in radar during World War II, the mature Dunbar chose to leave the military and pursue a civilian career in electrical engineering.

So can we claim Dunbar's twentieth-century flirtation with military science and engineering as evidence that this was the same person who had been a military scientist/engineer in previous existences? Not really. It seems that Dunbar's decisions in this latest incarnation were much influenced by his perusals of the readings Cayce gave him when he was a child. And, assuming his parents followed Cayce's orders, no doubt parental guidance also shaped his

➤ ANDIVIUS'S AMERICAN SPOKESMAN ⋘

The American novelist and poet Edward Lucas White based many of his stories on dreams, some of which seem to have been reincarnatory. His novel *Andivius Hedulio* (1921), the tale of a Roman nobleman, might appear to be a case of his being, as it were, merely a mouthpiece for one of his past selves. In an afterword to the novel, he explained:

> *The phrasing of this book is mine; otherwise I am scarcely more responsible for it than would be a secretary who had written it out from dictation. I did not originate the plot; I did not, except in a very few [cases], invent the scenes, incidents, or episodes, or create the characters. I dreamed the entire story . . .*

The tale contains an impressive amount of historical detail that, White found on checking later, was largely valid. This seems less remarkable when we bear in mind that throughout his professional career White was a teacher at Baltimore's University School for Boys, where the subjects he taught included classical languages. The best guess is that he subconsciously pieced together his knowledge of Roman life—more extensive than most of us could boast, because of his profession—and constructed his novel on that foundation.

later life. Yet this case is regarded as one of the most strongly supportive demonstrations of the veracity of Cayce's reincarnation readings.

Cayce told his printer friend, Arthur Lammers, that Lammers had been a Spanish monk in a previous lifetime. Over the next few weeks, under Lammers's friendly pressure, Cayce discovered others of the printer's previous existences. It was a couple of weeks after

Cayce's first "life reading" for Lammers that Cayce's unconscious revealed that the two men had been at Troy together. It seems to have been Lammers who first voiced the notion that Cayce might discover other such reincarnatory links—that souls must in some way tend to gravitate toward the same people in existence after existence. Not surprisingly, Cayce found plenty of "evidence" to support this hypothesis as he carried out his "life readings" for a wider and wider circle.

According to Jess Stearn's *Intimates through Time* (1989), Cayce was able to identify six of his own prior existences:

1. Asule, the "male half of a twin soul" in Atlantis.
2. Ra-Ta, an Egyptian high priest forced into exile because of an adulterous affair.
3. Uhjltd (pronounce that!), a Persian tribal chief who possessed psychic healing powers.
4. Xenon, the Trojan gatekeeper who, corrupted by his mistress, allowed the wooden horse to be brought into the city.
5. Lucius of Cyrene, a minor disciple of Christ mentioned in the New Testament.
6. John Bainbridge, an English gambler and soldier of fortune who became a scout in colonial Virginia.

Most of these existences were strongly marked by an inability to resist the temptations of the flesh. Stearn describes a visit to the Cayce Foundation many years after Cayce's death when, amid what seem to have been scenes of giggling and blushing, he discovered that several of the women now working in various capacities there had been Cayce's paramours in previous lives—not, Cayce's onetime secretary stressed to Stearn, in his most recent life, as by then he had finally gotten his lusts under control. It was because

The Procession of the Trojan Horse in Troy (1773) by Italian painter Giovanni Domenico Tiepolo. In one of his incarnations, Edgar Cayce purportedly was the corrupt guard who let the wooden horse into the legendary city.

of the reincarnatory magnetism that Lammers had postulated that the women had come to be associated with Cayce in the twentieth century.

XENOGLOSSY

If it were possible to find potential reincarnates who had the ability to speak in the language prevalent in the environment of their previous incarnations—a language they had no possibility of knowing how to speak in their current existence—would that not provide powerful evidence for reincarnation? Thus reasoned (quite correctly) Ian Stevenson, who devoted two books to the subject: *Xenoglossy* (1974) and *Unlearned Language* (1984). In these two books, Stevenson presents a total of three cases of regressed hypnotic subjects displaying the ability to speak or write a language they couldn't have learned— an ability to which Charles Richet gave the name *xenoglossy*:

1. "Jensen Jacoby," putatively a seventeenth-century Swede who represented a past incarnation of a thirty-seven-year-old American housewife called "TE"; the hypnotist was TE's husband.

2. "Gretchen," putatively a German who represented a past incarnation of American housewife Dolores Jay; the hypnotist was Jay's husband.

3. "Sharada," putatively a nineteenth-century Bengali who represented a past incarnation of a Maharashtra woman in her thirties named Uttara Huddar. There was no hypnotist in this instance; Huddar's "past incarnation" emerged while she was being treated for mental illness.

All three cases were investigated as deeply as possible by Stevenson to rule out fraud or hoax. Where Stevenson's research was less exhaustive, however, was in regard to the subjects' usage of the foreign languages' syntax. Independently, Sarah Grey Thomason, a professor of linguistics at the University of Pittsburgh, researched the three cases (and more), publishing her results in a series of papers during the 1980s that she summarized in a long article, "Xenoglossy," for Gordon Stein's *The Encyclopedia of the Paranormal* (1996).

Thomason devotes her major discussion to the first of these personalities, Jensen Jacoby, since the other two cases show similarities that need not be repeated. Although those who witnessed TE's hypnotic regressions were convinced that Jensen spoke the language, Thomason discovered that (a) Jensen's Swedish vocabulary included only about a hundred words (with some Norwegian words thrown in), (b) his pronunciation was often curious, and (c) he rarely uttered a complete sentence. And occasionally he got something completely wrong: Once, on being asked the price of some market item, he answered, "My wife."

Of course, that's still a hundred words more of Swedish than most non-Swedes speak! However, bear in mind the identity of TE's hypnotist: her husband. Hypnotic subjects are characteristically eager to please their interrogators, and it seems perfectly likely that TE's unconscious reckoned it would please her spouse if she came out with some Swedish. Although the conscious TE was convinced she knew nothing of the language, it's quite possible the *un*conscious TE— whose parents also spoke Polish, Russian, and Yiddish in the home— picked up cognates or bits and pieces of the language from her multilingual environment. For example, simply watching an Ingmar Bergman movie could provide a significant boost to one's Swedish vocabulary, thanks to the approximate translations in the subtitles.

The Gretchen case was very similar, although the subject, Dolores Jay, admitted openly that she had spent a little while studying a German dictionary after her husband had started the regression sessions. Even so, as Stevenson pointed out, before then she had already produced over two hundred German words while under her husband's hypnosis. An immediate point is that, in most American communities, one is far more likely to come across occasional snatches of German than of Swedish. Furthermore, as Thomason points out, the vast majority of Gretchen's responses to questions consisted of just one or two words, and quite a few were basically repetitions of the question with the intonation shifted from interrogative to declarative. Again, there were oddities of pronunciation and mistakes in basic vocabulary (e.g., *Bettzimmer* rather than *Schlafzimmer* for the common word "bedroom").

The Sharada case is somewhat different, occurring not in the United States but in India, and first making itself evident not under hypnotic regression but during psychiatric treatment. It also occurred in a society where belief in reincarnation is the norm— which means people are likely to be looking for indications of reincarnation, rather than for reasons to doubt it. The subject, Uttara

⇒ FROM INDIA TO THE PLANET MARS ⇐

Daughter of Jairus (1913) by Hélène Smith.

Catherine-Elise Müller, known to the world as Hélène Smith, was a French psychic who became famous with the publication of *Des Indes à la Planete Mars* (1899; translated as *From India to the Planet Mars*) by the distinguished Swiss psychologist and parapsychologist Théodore Flournoy. When Smith went into a trance, she would come under the control of a spirit guide called Léopold. (An earlier guide had been the spirit of Victor Hugo.) She would then take on the characteristics of various historical personalities, whom it was assumed represented her previous incarnations. One was Marie Antoinette; another was a princess called Simandini, one of the wives of Prince Sivrouka Nayaka of India; yet another was a Martian woman. Smith described the Martian environment and the people there in some detail, including an individual named Astané, whom Flournoy thought might be intended as a sort of Martian avatar of Léopold.

And then there's Astané's scary pet:

> . . . which caused Hélène much fright on account of its grotesque form—about two feet long, with a flat tail; it has the "head of a cabbage," with a big green eye in the middle (like the eye of a peacock feather), and five or six pairs of paws, or ears all about. . . . This animal unites the intelligence of the dog with the stupidity of the parrot, since on the one hand it obeys Astané and fetches objects at his command (we do not know how), while, on the other hand, it knows how to write, but in a manner purely mechanical. (We have never had a specimen of this handwriting.)

Flournoy was unimpressed by the notion that Smith's previous incarnations were speaking through her. In the persona of Simandini, she spoke in a strange language that Léopold claimed was "Ancient Hindoo" or Sanskrit, while her Martian persona brought forth a different language—naturally identified as "Martian." Flournoy brought in Ferdinand de Saussure, a specialist in Eastern languages, to analyze the "Sanskrit" and found that, whatever it was, it certainly wasn't a Hindu language. Flournoy concluded that both it and "Martian" were examples of "glossolalia," the nonsensical but language-resembling "speaking in tongues" sometimes displayed by people in ecstatic trances.

He diagnosed Smith's case as a fascinating example of what we now call dissociative identity disorder, and he thought that perhaps there might be some telepathy involved, too. Flournoy also found evidence that Smith's "previous incarnations" were products not of the past but of sources in Smith's current life. For example, she gave quite a few seemingly convincing details of her experiences during a "past life" as an Arab princess—details that Flournoy discovered came from a book Smith had once read (moreover, where the book made mistakes, so

Ferdinand de Saussure, the linguist who exploded Hélène Smith's claims to be speaking Sanskrit.

did Smith's "recollections"). Again, "Marie Antoinette" smoked cigarettes until it was pointed out to her that cigarettes hadn't been invented when Marie Antoinette was alive, at which point she promptly kicked the habit.

Smith was not pleased when Flournoy's book came out and she discovered he was dismissing her spiritual experiences as either products of a mental condition or, in effect, self-indulgent delusions. As might be expected, she broke off contact with him.

Huddar, had received a few lessons in Bengali, had read translated versions of Bengali novels, and had studied Sanskrit, so the fact that her previous incarnation, Sharada, could produce at least a passable rendition of Bengali was not so improbable. Stevenson found a couple of Bengali speakers who said Sharada's Bengali was perfectly acceptable, a couple who said it sounded like someone using Bengali as a second language, and a Bengali linguist who essentially ridiculed Sharada's use of the language from every conceivable angle. Stevenson tried to explain this latter judgment away along the lines that the linguist had spent only a short time with Sharada—not long enough for her to "warm up" to him.

In all three cases, there's the matter of pronunciation. One argument used by Stevenson and others is that someone who has to speak through the mouth and vocal cords of another might find occasional difficulty in attaining the desired pronunciation. However, Thomason points out that in Stevenson's first two cases—the only two worth much analysis—the supposed past incarnations showed a greater mastery, however slight, of their languages in speaking than they did in understanding what was said to them. With natural users of languages, this balance is just the opposite: Our passive vocabulary is much broader than our active one.

In her article, she also analyzes three cases provided to her by the Pittsburgh hypnotist Ralph Grossi and finds them even more lacking. In the first, a woman believed to be speaking Bulgarian while under regression was in fact speaking no extant language at all—just making appropriately Slavic noises and stringing them together in a semblance of grammar. The second subject was supposed to be a fourteenth-century Normandy knight speaking Gaelic; what he was really speaking was a sort of bastardized version of Church Latin treated to an Inspector Clouseau–style Frenchification. The third subject supposedly had as her previous incarnation an Apache woman who declined to speak any Apache but instead employed a form of Pidgin English that would have made even Tonto's scriptwriters gag. All three cases seemed obviously bogus—even though, to reiterate, there's not the slightest suspicion that anyone was acting dishonestly.

❧ PAST-LIFE REGRESSION HYPNOSIS

It's a central tenet among psychologists who perform past-life hypnotic regressions that they're doing this not merely out of morbid curiosity or in hopes of writing a best seller, but because it's actually good for the patients. Just as bringing to the surface and confronting one's past traumas in the current life can be psychologically helpful, so, the reasoning goes, can doing the same with the traumas of previous lives. At the very least, one might gain an understanding of one's current phobias or other psychological peccadilloes. Furthermore, even if it turns out that the patients aren't really remembering their prior existences but are instead constructing fantasies, surely those fantasies and the act of constructing them are likely to be therapeutic in the same way that psychodrama can be, allowing patients to voice matters they'd be ashamed or frightened to admit about themselves but can readily attribute to a "former self." (Recall that Joan Grant and Denys Kelsey came to a similar conclusion in their work.)

Nevertheless, numerous psychotherapists have spoken out strongly concerning the dangers of past-life regression. The "memories" induced by the practice tend to be as vivid as real-life experiences and quite as capable of inflicting psychological traumas that later have to be worked out. And even Ian Stevenson, that great champion of reincarnation, expressed great doubts about any possibility of past-life regression having therapeutic value. As he pointed out, many of the hundreds of children he interviewed had vivid "memories" of some truly horrific traumas their former incarnations had experienced. Rather than helping them overcome their current-life anxieties and phobias, such recollections very often reflected and even exacerbated those anxieties. Someone who had allegedly died in a car crash in a previous life, for example, was likely to be absurdly wary of cars, roads, and traffic in their present existence.

Leaving aside issues of therapy, many people naturally regard the results of past-life regressions as straightforward proof of the phenomenon of reincarnation. In reality, this is a very difficult case to argue—for a number of reasons. There's a lot of extremely convincing evidence that subjects invent stories of past lives to please the hypnotist (as also happens notoriously in "false memory syndrome"). Likewise, it's generally evident on reading the transcripts of these sessions that the hypnotist is, consciously or unconsciously, leading the patient on—conveying through countless hints what it is that the hypnotist wishes to hear. (Your fifteenth birthday party? Bo-*ring*. Your life as a nymphomaniacal Atlantean warrior queen? Now *that's* more like it!)

BRIDEY MURPHY

The technique of asking hypnotic subjects to cast their recollections back beyond birth and recall past lives owes its modern popularity to the Bridey Murphy case. This was a cause célèbre during the early 1950s, following sensationalist reports in the *Denver Post*. Colorado amateur hypnotist Morey Bernstein "regressed" Virginia Tighe (to whom he gave the pseudonym "Ruth Simmons"), the wife of an acquaintance, to a previous life as Bridey Murphy. Murphy had apparently been born in Cork, Ireland, in 1798, latterly living in Belfast, where she died in 1864. Bernstein's book, *The Search for Bridey Murphy* (1952), was a runaway best seller and sparked, among much else, a brief fad for Bridey Murphy fancy-dress come-as-you-*were* parties. Oddly, the guests tended to turn up dressed not drably, as a peasant like Bridey might have, but as Cleopatra, Grigory Rasputin, Elizabeth I, and the like. The case inspired several pop songs, a cocktail, even a dance.

But then the *Chicago American* published a series of revelations about the circumstances underlying the case: Tighe's aunt, Mary Burns, was "as Irish as the lakes of Killarney" and had told her stories of her Irish youth when Tighe was an infant. Later, during Tighe's childhood, she had had a neighbor named Bridey Corkell, née Murphy, and had been captivated by both Corkell's Irish background and her son, John Corkell. This all seemed like it was enough to explain everything, and the reputation of the Bridey Murphy case plummeted.

The trouble was that the *Chicago American*'s "revelations" didn't check out. For example, far from being "as Irish as the lakes of Killarney," Mary Burns was a native of New York. Far from telling the infant Virginia tales of Ireland, Burns didn't even meet her niece until the girl was eighteen. Undisclosed by the *Chicago American*, John Corkell was an employee of the newspaper; Tighe had barely known him and his mother. And the *Chicago American* had been mightily peeved when it failed to purchase the serial rights to Bernstein's book, which had gone instead to the rival *Chicago Daily News*. So the exposé itself was as dubious as Bernstein's claims.

More authentic problems with the story were unearthed by William J. Barker, the reporter who had written the original articles in the *Denver Post*. The *Post* sent him to Ireland to dig up concrete evidence of Bridey's former life, and he found none—just discrepancies. Bridey had supposedly lived in the early 1800s in Cork in a wooden house and slept in a metal bed, yet houses there in the early 1800s were made almost exclusively of brick or stone and the first iron bedstead didn't reach Ireland until about 1850. Bridey's husband had supposedly taught at the law school of Queen's University in Belfast. Yet this university, originally known as Queen's College, Belfast, wasn't chartered until 1845, didn't have a law school, and didn't become Queen's University until 1908. The church she supposedly attended wasn't built until 1911. And Bridey's "Irish" speech, so

Queen's College, Belfast, depicted in 1851, just a few years after it was built.

convincing to American ears, sounded just laughable to Irish ones. Americanisms and mispronounced words abounded—she even pronounced her husband's name, Sean, as "See-ann"!

Such concerns didn't undermine the success of the 1956 movie based on Bernstein's book, *The Search for Bridey Murphy*, starring Teresa Wright. But the case brought some associated tragedies. When nineteen-year-old newsboy Richard Swink of Shawnee, Oklahoma, shot himself in the mid-1950s, he left a suicide note, saying, "I am curious about this Bridey Murphy story, so I am going to investigate the theory in person."

The Bridey Murphy case wasn't the first example of supposed past-life regression. In the relatively early days of the craze for L. Ron Hubbard's Dianetics, the practice of auditing by means of an Electropsychometer, or E-meter, was used to elicit information not only about a subject's "engrams" (subconscious mental images of traumatic experiences) but also about the subject's past lives. It was this "discovery" that led Hubbard to transform Dianetics into the religion of Scientology, which holds reincarnation as a central tenet of its whole science-fictional mishmash.

MANY LIVES, MANY QUESTIONS

Another case of memories of past lives apparently being disinterred by use of hypnosis is recounted by Brian L. Weiss in his book *Many*

Teresa Wright, star of the Hollywood movie *The Search for Bridey Murphy* (1956).

Lives, Many Masters (1988). Weiss was a psychiatrist to whom a patient called "Catherine" was referred in 1980. He attempted to cure her multiple anxieties by regressing her back to childhood. Suddenly, during a session in 1982, she began to talk about being an eighteen-year-old Egyptian named Aronda who had died by drowning in 1863 BCE. Since one of Catherine's phobias had been about choking and drowning, and since this abated considerably after her past-life regression, Weiss repeated the process.[6]

In the course of further sessions, Catherine explored previous existences as a fifteenth-century Dutchman, an eighteenth-century Ukrainian, a nineteenth-century Virginian, and so on. In fact, she proved to have had no fewer than eighty-six incarnations, many of

---

[6] Weiss states as if it were unremarkable that both Catherine and himself became increasingly psychic as the sessions continued.

which she described to the agog Weiss. In many of these past lives, individuals from her current life (Weiss included) appeared in different roles. She also experienced odd intermissions between the incarnations in which she seemed to be in deep communion with spiritual masters who were giving her advice to bring with her into the next lifetime:

> *Our task is to learn, to become God-like through knowledge. We know so little. You are here to be my teacher. I have so much to learn. By knowledge we approach God, and then we can rest. Then we come back to teach and help others.*

As Weiss points out, it was as if he were witnessing a practical demonstration of the truth of the Tibetan Book of the Dead—the eighth-century text describing the experiences of consciousness between death and the next rebirth—and therein lies the probable explanation of what was going on. There's every possibility that Catherine had read, or read about, the Tibetan Book of the Dead (it's easy enough to find a copy).

And there were some oddities about the whole affair. During that very first regression, Aronda had been good enough to give Weiss the date: 1863 BCE. How could any ancient Egyptian have known to use a dating system based on an event nearly two millennia in that person's future? Of course, one could claim Aronda was speaking through the modern-day Catherine, and that the latter, even in a trance, was able to "translate" from the ancient Egyptian dating system to one that Weiss might comprehend, but this seems like a bit of a stretch.

Another concern is that at least two of Catherine's previous incarnations overlap: How could she have been both a Ukrainian boy and a Spanish prostitute at the same time? Quizzed by journalist Tom Shroder on these and other points,[7] Weiss responded, "The totality of

---

[7] Tom Shroder, *Old Souls* (1999).

the experience was such that these inconsistencies only add to its complexity. There is so much we don't know." Shroder adds that Weiss,

> *. . . years after* Many Lives, Many Masters *. . . distanced himself from the idea that regressions proved the reality of reincarnation. What he cared about, he declared, was that whatever these regressions tapped into, even if only the patient's subconscious, had proved to be tremendously helpful in therapy. He had seen problems resistant to all other kinds of treatment clear up almost instantaneously after dramatic regressions. I asked him if he had done any clinical studies to verify his impression that regression therapy got such dramatic results. He hadn't, he said, but he wished that somebody would.*

THE BLOXHAM TAPES

"Jane Evans" was the pseudonym of a Welsh housewife who was one of the subjects of a series of purported past-life regressions conducted over a period of twenty years or more by hypnotherapist Arnall Bloxham. The regression transcripts were serialized in the London *Sunday Times*. They were also the subject of a BBC TV documentary, *The Bloxham Tapes* (1976), presented by Magnus Magnusson. That same year the program's producer, Jeffrey Iverson, published the book *More Lives Than One?* based on the documentary. Iverson was particularly fascinated by the Evans case, and for a while it was something of a cause célèbre.

Bloxham had good credentials as a hypnotherapist—indeed, he had served for a time as president of the British Society of Hypnotherapists—but he was not a historian. Though neither he nor Iverson realized it, this fact severely compromised the supposed veracity of the reincarnation evidence he produced through his egressions. It should also be noted that Iverson's was not the first book about Bloxham's past-life regression work; in 1958 Arnall's wife,

Dulcie Bloxham, had published *Who Was Ann Ockenden?* Ockenden was a schoolteacher who, under Bloxham's hypnosis, seemed to recall seven previous existences.

Jane Evans recalled six. In one of these she had been a maid in the household of the French merchant Jacques Coeur (c. 1395–1456), whom she described as a bachelor, and under hypnosis she recounted all sorts of details about his home. The latter feat becomes a little less extraordinary when we discover that photographs of Coeur's splendid mansion in Bourges have been very widely published. And her description of him as unmarried and childless seems odd; the real-life Coeur was married and had five children. The solution to this puzzle seems to be that Evans had read *The Moneyman* (1947), Thomas B. Costain's popular novel based on Coeur's life.[8]

And a novel seems to lie at the heart of another of Evans's "past lives." She supplied a startling amount of detail about her existence as a woman named Livonia, who lived in the household of Constantius, father of Constantine the Great, during the Roman occupation of Britain. Upon checking, much of this detail proved to be highly accurate, but a mystery remained as to who Livonia might have been. That there was no record of her was perhaps not so surprising—attitudes toward women have frequently relegated them to the shadows over the centuries. But one might have expected some reference to turn up of Marcus Favonius Facilis, a first-century Roman centurion whom Evans described as Constantine's military tutor. Quite a while passed before someone linked the two characters of Evans's "memories" to the novel *The Living Wood* (1947) by Louis de Wohl, which was

---

[8] In "The Bloxham Tapes Revisited" (*Journal of Regression Therapy*, July 2008) Ian Lawton suggests quite forcefully that there must have been more to it than this, since Evans produced plenty of other historical details that are still relatively obscure and that aren't mentioned in Costain's novel. Yet Lawton has difficulty getting around the marital and paternal discrepancies.

The Cour du Palais in Bourges, France: the stately home of the merchant Jacques Coeur.

set in Roman-occupied England and contained invented characters including Livonia and the military tutor Marcus Favonius Facilis. Other minor characters de Wohl had created—Titus, Curio, and Valerius—also appear, in the same roles, in the Evans transcripts.

In yet another existence, Evans was a Jewish woman named Rebecca living in twelfth-century York. The locals were anti-Semitic, and Rebecca and her family were forced to wear a yellow circle over their hearts. One day there was a pogrom, and the family fled, first hiding in the castle and then, when that seemed likely soon to be overrun by the mob, in the crypt of a local church. While Rebecca's husband and son were out trying to scavenge some food, the mob found Rebecca and her daughter in the crypt and murdered them. Historians determined that the church described by Evans must have been St. Mary's, Castlegate, except that unfortunately St. Mary's didn't have a crypt. Much has been made of the fact that in 1975, many years after Evans's sessions with Bloxham, excavations at St. Mary's revealed that the church seemingly had indeed once had a crypt, long ago blocked off and forgotten. Far

Tomb of the Roman centurion Marcus Favonius Facilis, dating to the first century CE, at Colchester Castle in England: de Wohl's inspiration for Constantine's military tutor character.

less widely reported is that over the succeeding decade it became clear that what had at first been thought to be a crypt was actually a charnel house, built some time after the Middle Ages.

Nonetheless, Barrie Dobson, professor of medieval history at the University of York, whom Iverson had consulted during the preparation of both the TV documentary and the book, identified other details mentioned by Evans that seemed like things only professional historians might know. Dobson was also perfectly prepared to accept that the Jews of York might have been forced to wear yellow circles in 1190, even though (a) it was not until 1215 that a papal decree was promulgated demanding that Jews be identified as such, and (b) outside Germany and France the yellow circle is not known to have been used to identify Jews.[9]

The solution here may lie in an altogether different piece of forgotten history. In *The After Death Experience*, Ian Wilson mentions that several sources have told him that they recall hearing a BBC radio play in the 1950s about the York Riots; alas, no one could supply him with any details. And of course a number of nonfictional accounts of the York Riots have been published, any one of which Evans might have come across. Another popular suggestion is that Jane Evans's Rebecca of York could be based on the character of that name in Sir Walter Scott's novel *Ivanhoe* (1820). This seems a dud hypothesis, though: There are scant similarities between the two Rebeccas. At best, it's possible that Evans's unconscious picked up the name from the novel—or from its 1952 screen adaptation—and contrived a biography to go with it.

No one has suggested that any of the principals involved in the Jane Evans case were guilty of the slightest dishonesty or lack of integrity. What appears to have happened once again is that Evans, like so many hypnotic subjects, tried to please her hypnotist. Bloxham, earnestly attempting to unearth her memories of previous existences,

---

[9] In York, the badge used after 1215 was apparently made up of white stripes.

unwittingly encouraged her entranced mind to fantasize freely, using material she had at some time either read, watched, or listened to— but which had long since vanished from her conscious memory—as the basis for her accounts.

DESTINY OF SOULS

Regression hypnotist Michael Newton focuses less on examining his patients' past lives and more on what happens to souls during the intervals between incarnations. In *Journey of Souls* (1994) and more expansively in *Destiny of Souls* (2000), he offers case studies wherein hypnotic subjects describe their adventures during those interludes. Like so many currently operating on the fringes of accepted knowledge, Newton claims that he began as a skeptic but his attitude changed when one day he accidentally regressed a patient not just to childhood but back through it and out the other side. As he reflects in *Destiny of Souls*:

> *Thus, when I unintentionally opened the gateway to the spirit world with a client, I was stunned. It seemed to me that most past-life regressionists thought our life between lives was just a hazy limbo that only served as a bridge from one past life to the next.*

From his descriptions, the abode of souls sounds sadly mundane. Although souls are born as particles of energy from amorphous clouds, they seem able to adopt human form and, indeed, to prefer that guise.[10] And their institutions seem depressingly earthlike, too:

---

[10] In Newton's afterlife, as in every other one I've come across, there are no souls of extraterrestrial species. This seems odd: If the astral plane is infinite and boundless, surely human souls should be encountering ETs all the time. Perhaps the explanation is that there *are* no extraterrestrial intelligences, or that there are but that ETs don't have souls. Or . . .

This twixt-life venue is in effect a high school in which the students have a certain amount of freedom but are under the thumb of the teachers (or Elders) and must follow the syllabus. Not that it's all work and no play. In *Destiny of Souls* Newton quotes one of his patients describing the reception he was given by his pals, the last time he returned to the abode of souls at the end of an earthly life:

> *After my last life, my group organized one hell of a party with music, wine, dancing and singing. They arranged everything to look like a classical Roman festival with marble halls, togas and all the exotic furnishings prevalent in our many lives together in the ancient world. Melissa (a primary soulmate) was waiting for me right up front, re-creating the age that I remember her best and looking as radiant as ever.*

And presumably in the afterlife there's no need for birth control!

After the postlife shindig, it's time for the soul to spend a while with teacher-guides, who offer instruction in things like what went wrong last time and how to plan for a better outcome next time. There's also plenty of allowance for fun group activities with one's own cluster of fellow souls, although this frantic socialization isn't compulsory: "If souls want solitude they can have it." You might well be required to have a couple of sessions with the council of Elders—souls at a higher level than the mere teacher-guides but not so exalted as the Presence. None of Newton's patients have ever actually encountered the Presence face-on, so to speak, but several have told how it lives in "a sphere of dense purple light." That mention of the color purple is no accident, because, according to Newton, "[E]nergy colors displayed by souls in the spirit world . . . relate to a soul's state of advancement."

All the parties and study periods finally done, it's time for you to take another sojourn back in physical form. You're "escorted to the space of embarkation for the trip to Earth." If this seems a bit *Star*

⇛ FUTURE-LIFE PROGRESSION ⇚

Another hypnotic regressionist who had his day in the sun was H. N. Banerjee, author of the books *Once and Future Life: An Astonishing Twenty-Five-Year Study on Reincarnation* (1979) and *Americans Who Have Been Reincarnated* (1980). Banerjee was very popular among movie stars, for reasons that aren't too hard to deduce: He helped them discover they'd had starring roles in their previous existences, too! But, if hypnotists can regress people to their past lives, there's an obvious question: Should it not be possible to progress people forward into their future lives? Sure enough, this has been done—a notable practitioner being Helen Wambach. She was able to dispatch a number of hypnotic subjects to the twenty-second century and beyond, where they described a planet slowly recovering from nuclear war and pollution. The astute reader will notice that these reports are from sufficiently far in the future that none of us could check them for accuracy.

However, some of the progressions were to lives in the nearer future, such as those recounted in *Mass Dreams of the Future* (1989) by Chet B. Snow, who met Wambach in 1983 and soon agreed to be progressed by her. His first journey was to the Arizona desert in mid-1998, where he found himself in a small community largely cut off from the world and eking out a subsistence living. Clearly, the world had been subject to some kind of ecocatastrophe because the sky was black at noon and earthquakes were common.

A few months later he made another trip to the future, this time touching down at Christmastime, 1996. He found that the ecocatastrophe hadn't happened yet, although the weather was beginning to become extreme and unpredictable. The world economy was plummeting in the wake of a stock market crash. Turning on the TV news, he discovered that there was trouble in the Middle East—hardly a predictive coup here—and that the current US president was "someone much younger than Ronald Reagan and he had more prominent ears. The face seemed vaguely familiar . . . I got the feeling that this man had been a United States

Senator or prominent governor in the 1980s but not someone I'd have predicted as President a dozen years later." Of course, "much younger than Ronald Reagan" hardly narrows down the field much. The bit about "prominent ears" doesn't seem indicative of Bill Clinton, and, although the rest of his description could apply to Clinton, it could also apply to most US presidential candidates. The stock market crash and the growing instability of the weather are features of today's world, but were not as much of a reality in 1996.

Wambach then jumped Snow forward to late 1997, when life was starting to get a bit grim in the Arizona community, and after that to late 1998, by which time society as we know it had largely collapsed (as had Japan and most of the US West Coast, having succumbed to tsunamis after monumental activity all around the Ring of Fire). The Russians[1] were in the process of conquering Europe, and the United States was under martial law.

Cover of a 1947 booklet published by the Catholic Catechetical Guild Educational Society, raising the specter of a Communist takeover.

Snow's next stop was at the end of 2000, where matters had hugely improved! The global environment seemed to be recovering at lightning speed. The social structure of the United States had disintegrated, but a new one was emerging, phoenixlike, thanks to the efforts of telepathic New Agers. And at his final stop, two years later, he found that new social allegiances had been forged as he rode north to visit comrades in now-balmy Alberta.

Those of us who lived through the years 1996–2000 may be wondering if Snow's accounts are entirely accurate.

---

[1] Snow refers to them as the "Soviets," so perhaps there had been another Bolshevik Revolution.

Wars, what happens next is somehow more Woody Allen: "Souls join their assigned hosts in the womb of the baby's mother sometime after the third month of pregnancy so they will have a sufficiently evolved brain to work with before term." This is necessary because they have to accustom themselves to once more operating with physical "brain circuitry." This means that the fetus, after the third month, is significantly more intelligent than the newborn child because it possesses all the memories and wisdom of an old soul. At the moment of birth, though, "an amnesiac block sets in."

❧ CRYPTOMNESIA

But how, one might ask, do the subjects produce such a wealth of historical detail[11] in describing their previous incarnations? In *Hidden Memories: Voices and Visions from Within* (1992), psychologist Robert A. Baker proposes that supposed recollections of prior lives are the product of cryptomnesia and what I've been calling narrativization.

The phenomenon of recalling long-lost memories as if they were fresh thoughts, *cryptomnesia* was first named and described by Théodore Flournoy in the context of the Hélène Smith case (page 198). Our brains store far more memories than we generally realize because many of these memories are not immediately available to us. On occasion, however, these memories can pop back into our conscious minds. When they do so, chances are that we won't recognize them as memories—instead believing them to be original thoughts—or we may accept that they are memories but misidentify how the memory was formed. This seems to be what happened to Jane Evans, for example, when her memories of reading a Louis de Wohl novel suddenly reappeared in her consciousness and were

---

[11] These historical details often prove correct, but are also frequently wrong, as we've seen. For obvious reasons the "misses" less commonly make it into the literature.

identified as memories of genuine past-life experiences. Add to that our innate tendency to link observations and memories together in such a way that they comprise a narrative, unconsciously filling in any gaps in the story with inventions of our own, and we see why eyewitness testimony is so notoriously unreliable and also why our anecdotes grow better with each retelling. That elderly uncle is genuinely unaware that his rambling anecdote bears little resemblance to what actually happened, having been substantially embellished over the years. We remember the best story.

But what about the sensory element? Jane Evans didn't "remember" her Roman lifetime just as dry words on the page; she "recalled" the sights and sounds and everything else, including the laughter of children and the brush of the breeze against her cheek. Here Baker refers to hallucinations in the sense that we can remember images and other sensory experiences that we've imagined—as, perhaps, when gripped by a Louis de Wohl novel—as if they were real. Our dreams may have been influenced by that novel at the time, and elements from those dreams may reappear in our consciousness. Again, our brain narrativizes all the information and creates from it a coherent pseudo-memory.

Experiments done on past-life regression in the 1950s and 1960s by two psychologists, one in the United States and one in Finland, demonstrated the perils of taking these "memories" at face value. In the United States, Edwin Zolik hypnotized volunteers and tape recorded their accounts of past lives. Later, Zolik played the tapes to the subjects. He then hypnotized them again, this time asking if they knew where the "memories" had come from. Almost without fail, they easily recalled mundane rather than supernatural sources. For example, in the first session, one subject clearly recalled having been an embittered old riverman who had died in 1878. But, when hypnotized a second time, the "grumpy riverman" was able to recognize that he'd created his account based on a character in a movie he'd once seen.

A few years later, in Finland, Reima Kampman performed a roughly similar experiment. One outstanding example concerned a teenage girl who, under hypnosis, recalled being an innkeeper's daughter named Dorothy in medieval England; she even sang the Middle English song "Sumer is Icumen In." When pressed under a second hypnosis to identify the source of this "past life," the subject volunteered that it came from a book she had once skimmed in a library. The song came from *The Story of Music* by British composers Benjamin Britten and Imogen Holst.

Reima Kampman was the inspiration behind the song "Mennyt Mies" (2012) by J. Karjalainen, the "Finnish Bruce Springsteen."

✸ A THEORETICAL FRAMEWORK?

Reincarnation is a popular explanation for déjà vu, that sudden "I've been here before" feeling. It also offers a pat explanation for the phenomenon of love at first sight, as Elisabeth Kübler-Ross pointed out:

> *Many people have had encounters with a fellow-man who seemed to them so familiar, as if they had known each other for decades and not just for a brief moment. And they often say jokingly, "Maybe we have been together in another lifetime." If they only knew how true this is in many cases . . .*[12]

---

[12] Foreword to *Reincarnation: The Phoenix Fire Mystery* (1961), edited by Joseph Head and S. L. Cranston.

American spiritual teacher Gary Zukav explains in his book *The Seat of the Soul* (1989) that people with a mere five senses, like most of us, are unaware of their other incarnations, but if through spiritual awakening we turn ourselves into "multisensory personalities," we can become so. In Zukav's view, the personality and the body are unimportant vehicles of the immortal soul.

I imagined the body as being rather like a rental car that you pick up at one depot, drive around for a while (a lifetime, in fact), and then leave at a different depot. While the car may be excellent and you may enjoy driving it, at no stage is it actually you: it's just something to get around in. After it has dropped the car off at the depot, the soul goes back to its natural state of immortality and timelessness.

Zukav's thesis is that our five senses have been ample so far to empower our explorations of the physical universe, but that, as we evolve into multisensory beings, we'll find new ways of apprehending the nature of the universe around us. When the book was published a quarter century ago, Zukav expected this great leap forward to happen imminently.

As of this writing, *The Seat of the Soul* is apparently Oprah Winfrey's all-time favorite book, aside from the Bible.

As an example of the sort of muddled thinking that lies behind most attempts to construct a theoretical framework for reincarnation, one could hardly find an explanation much more muddled than this, from American philosopher Francis Bowen's essay, "Christian Metempsychosis" (1881):

> *Our life upon earth is rightly held to be a discipline and a preparation for a higher and eternal life hereafter. But if limited to the duration of a single mortal body, it is so brief as to seem hardly sufficient for so grand a purpose. Threescore years and ten must surely be an inadequate preparation for eternity. But what assurance have we that the probation of the*

soul is confined within so narrow limits? Why may it not be continued, or repeated, through a long series of successive generations, the same personality animating one after another an indefinite number of tenements of flesh, and carrying forward into each the training it has received, the character it has formed, the temper and dispositions it has indulged, in the stage of existence immediately preceding? . . .

Why should it be thought incredible that the same soul should inhabit in succession an indefinite number of mortal bodies, and thus prolong its experience and its probation till it has become in every sense ripe for heaven or the final judgment? Even during this one life our bodies are perpetually changing, [though] by a process of decay and restoration which is so gradual that it escapes our notice. Every human being thus dwells successively in many bodies, even during one short life.

We're reminded of the Voltaire quip at the beginning of this chapter, except that Bowen is deadly serious in his belief that reincarnation (metempsychosis) is no more remarkable than cell regeneration—this even though, with minimal technological help, we can see cell regeneration happening while we have no clear demonstration that reincarnation is anything more than a fantasy.

Whatever the implausibility of reincarnation as a phenomenon, and however hard it might be to imagine a mechanism for it,[13] what seems most difficult to answer of all the questions surrounding reincarnation is this: Why? Evolution is not a purposive force. Because of natural selection, disadvantageous attributes tend to die out after a while, as do useless ones—although the latter are likely to hang around longer, or even to mutate into quite different attributes that

---

[13] Let me qualify that: All sorts of people have demonstrated that it's perfectly easy to imagine a mechanism for reincarnation. I'm talking about a *viable* mechanism.

do have a use. Could reincarnation have ever been a better than useless attribute of our species? Jane Roberts, author of the Seth books, said that yes, there *is* an evolutionary benefit to reincarnation, although the evolution concerned is of the soul, rather than of the body. Cultural evolution appears not to have entered her equation. And, in *The Seat of the Soul*, Zukav seems to agree: Human evolution appears, in his model, to be based upon "the continual incarnation and reincarnation of the energy of the soul into physical reality for the purposes of healing and balancing its energy in accordance with the law of karma." Of course, he's speaking about spiritual evolution rather than the physical version.

Conversely, it's not hard to find arguments pointing to reincarnation as being in fact *dis*advantageous: Just as all species rely for their physical advance on the oldsters dropping out when their time is up, so does our own culture's intellectual advance. The "wisdom of the ancients" has its uses, but just look at the stagnation in any culture that's ever allowed itself to become a gerontocracy. The USSR in the decades before the appointment of Mikhail Gorbachev is a recent example.

If concluding that reincarnation is at best mildly disadvantageous for a culture and by extension a species, one could claim that at some stage in our upward climb there was a use for reincarnation, although that use vanished many generations ago. Perhaps the process itself is—like that other mildly disadvantageous attribute of ours, the appendix—taking far longer to do likewise. But, even if we accept that as a possible argument, we're still left with the conundrum as to what evolutionary advantage there could ever have been in the old surviving for many times the physical lifespan.

And this leads to another question about reincarnation. Although doubtless there are plenty of brightly colored books claiming that domestic pets reincarnate, most serious reincarnationists, whether motivated by religion or by pseudoscience, concur that humans are

⇝ A BARGAIN MISSED ⇜

In his book *Old Souls* (1999), Tom Shroder mentions discovering the website for Fountain of Youth Reincarnation Systems, a company offering to sell you, for a mere $399, your very own kit to create a Happy Ever After Return Package (inaccurately acronymized as HARP). Using your HARP, your soul will, after death, be able to navigate its way safely back into a new body. Otherwise, well, who knows what might happen?

It's hard to know if this was a scam or a spoof. By the time I tried to check it out in early 2015, there was no trace online of the Fountain of Youth Reincarnation Systems website except some references to Shroder's book and, rather poignantly, a page with a dead link to it promoting the services of the web architect who designed the FoYRS site.

the only creatures capable of return. At what stage, then, during our evolutionary history did we acquire the ability to reincarnate? If one believes in the existence of souls, which presumably arrived at some point after the human animal had begun to distinguish itself from the other primates, it might be legitimate to link the two acquisitions. After all, most attempts to rationalize reincarnation assume that what is actually doing the transmigration is a soul. Or was it, perhaps, the arrival of the soul and/or the ability to reincarnate that was the very mutation that set us off on a divergent evolutionary course from that of our primate kin?

These are good questions for a science fiction writer to ponder, perhaps, but they seem to be in a different room from science. At the moment, in scientific terms, reincarnation fails on the grounds (among many others) that:

- ◆ There seems to be no evidence for it beyond the anecdotal.
- ◆ No one has yet come up with a mechanism for it that doesn't rely on yet further unknowns and wild conjectures.
- ◆ It seems to have no function or indeed to be actively disadvantageous in evolutionary terms.

POSTSCRIPT: BRAINROT

"If scientists were paid by results, the last parapsychologist would have long since died of hunger . . ."

—Alexander Baron, *Exploding the "Psychic Detective" Myth* (1993)

IN 1994, THE DUTCH PSYCHOLOGIST SYBO SCHOUTEN published in the *Journal of the American Society for Psychical Research* the conclusions of a five-year experiment, comparing results obtained from self-described psychics with those from a control group of ordinary jills and joes. Each individual was visited several times a year, shown a photograph of someone unknown to them, and asked what they thought the person in the photo was like. The researchers took over ten thousand statements and found there was no statistically significant difference in the success rates of the two groups.

When I told a friend I was working on a book about science's confrontations with—and often demolitions of—psychic claims, he looked at me incredulously and grunted something to the effect that surely no one much believed in that supernatural hogwash any longer, did they? So in early January 2015, as I neared the end of the book's writing, I was delighted to be able to send him this want ad from the Portland, Oregon, craigslist website:

Beginning Demonologist Looking for a Mentor
Greetings,

> *I am a fledgling Demonologist looking for an experienced*
> *mentor to take me under their wing. I have been doing a lot of*
> *studying and research and I feel like I am ready to serve God*
> *in the battle and help people become free from demonic*
> *infestation. I need field and investigation experience from a*
> *seasoned Demonologist or Paranormal Researcher. I have*
> *mainly studied Catholic based text like Adam [Blai] or [Ed*

and Lorraine Warren], however, I am starting to research other religious beliefs and their rituals.

Please, seriously experienced Demonologists or Paranormal Researchers only. I take this very seriously.

On November 12, 2014, the world watched as the European Space Association (ESA) craft *Rosetta*, at the end of a ten-year mission, made a rendezvous with comet 67P/Churyumov-Gerasimenko and sent down a smaller craft, *Philae*, for our species' first ever soft landing on a comet. The landing didn't go without a hitch, but it nevertheless represented an extraordinary achievement. A few months later, in July 2015, NASA's *New Horizons* probe went through the system of Pluto, giving us for the first time close-ups of this most popular minor planet. "I think the solar system saved the best for last," said project leader Alan Stern. There was a sense that these ventures had rekindled the world's excitement about science.

Around that time, the US Senate was appointing to the chairs of its two major science committees two men who are triumphalist in their denial of science: Ted Cruz and James Inhofe. It takes no genius to realize that the consequences of this are likely to be disastrous for the nation, in both the shorter term and the longer. By contrast with irrationalism on this scale, it might seem that belief in superstitions, ghosts, telepathy, the afterlife, and all the rest of the paraphernalia of the supernatural is somewhat inconsequential—indeed, harmless. But it contributes to what we can think of as a collective brainrot.

If we believe foolish things in one field of endeavor, we're all the more likely to believe foolish things in another. People often use the term "magical thinking" to describe many aspects of the supernatural and the paranormal. The "magical" part of it refers to the mistaken attribution of a particular cause to a particular effect. For example, if you cross your fingers before placing a bet and your horse comes in first, you might attribute the win to the crossing of the fingers. We hear a bump in the night and assume there's a spook about rather than that it was just the blasted cat knocking something over, yet again. I could produce further examples *ad infinitum*. We tend to reinforce the magical thinking by ignoring those times when the horse didn't come in first, and so on.

A special case of magical thinking is what I sometimes call the "Clap Your Hands if You Believe in Fairies Syndrome." The reference is to the emotional climax of J. M. Barrie's stage play *Peter Pan; or, The Boy Who Wouldn't Grow Up* (1904), in which Tinkerbell—seen only as a point of light throughout the play—is near to death, her light flickering and fading fast. The audience is told that the only way to save her is to reaffirm their belief in fairies by clapping their hands as loud as they can. And so lots of small children with tear-stained faces drop their popcorn all over the floor.

The grown-ups realize, even in the moment, that it's all just a fiction. They clap along, of course, but that's only for the sake of the kids. As adults, we realize that the existence or nonexistence of fairies is entirely unaffected by our belief in them. This seems so obvious to us that it's hardly worth mentioning. And yet, even though we should—and do—know better, we apply this form of magical thinking to all sorts of situations in everyday life. We fall into the habit of thinking that, if only we believe hard enough in something—whether it be passing that exam, gaining that promotion, getting past first base with the gal or guy of our dreams—we can make it happen. And, again, we reinforce this by remembering the times when it "works" and forgetting the times when it "doesn't."

Magical thinking can obviously lead us as individuals into some bad life decisions, but similar thinking at the level of crowds can be very dangerous, indeed. The obvious example on the national scale at the moment is the piece of magical thinking that holds that the problem of man-made climate change can be solved by simply pretending it doesn't exist—a technique that no one in their right minds would apply to, say, a broken leg, but which millions accept as a reasonable solution to a major global problem.

And, just in the same way that far too many students of the psychic have fallen into the trap of using special pleading or inventing extenuating circumstances when confronted by fraud or error, our magical thinkers on climate invent global conspiracies among climate scientists, inflate minor disagreements among said scientists or trivial details in the data to make them seem of epic importance, and display all the other characteristics we've come to recognize in our jaunt through the study of the supernatural—right down to the "just because *that* climate-denying politician has been shown to be in the pay of the fossil-fuels companies doesn't mean they all are" argument. In the same way that many students of the psychic have performed every possible mental contortion rather than admit what the evidence staring them in the face indicates, so do far too many of the magical thinkers strutting our national stage persuade themselves that the false is true—or, at least, pretend that they do so.

The truth is that it's not all about us. Reality is unaffected by whatever we might choose to believe about it. However hard we might wish that things were one way or the other, they stay the same unless we actively change them. We need to stop believing that reality will be other than what it is just because we would like it to be so, in the same way that, as we got older, we realized that all the clapping in the world wasn't going to make fairies exist.

We really do need to grow up.

BIBLIOGRAPHY

Abbott, David P. *Behind the Scenes with the Mediums.* Chicago: Open Court, 1907

Alexander, Eben. *Proof of Heaven: A Neurosurgeon's Journey into the Afterlife.* New York: Simon & Schuster, 2012

Alexander, Eben. "Heaven is Real: A Doctor's Experience with the Afterlife." *Newsweek,* October 8, 2012

Alexander, Eben. *The Map of Heaven: How Science, Religion, and Ordinary People Are Proving the Afterlife.* New York: Simon & Schuster, 2014

Anson, Jay. *The Amityville Horror.* New York: Bantam, 1979

Aveni, Anthony. *Behind the Crystal Ball: Magic, Science, and the Occult from Antiquity through the New Age.* New York: Times, 1996

Baker, Robert A. *Hidden Memories: Voices and Visions from Within.* Amherst: Prometheus, 1992

Barrett, J. O. *Looking Beyond: A Souvenir of Love to the Bereft of Every Home.* Boston: William White, 1871

Barrett, Sir William. *Death-Bed Visions: The Psychical Experiences of the Dying.* London: Methuen, 1926

Bernstein, Morey. *The Search for Bridey Murphy.* New York: Pocket, 1978

Blackmore, Susan J. *Beyond the Body: An Investigation of Out-of-the-Body Experiences,* 2nd edn. Chicago: Academy Chicago Publishers, 1992

Blum, Deborah. *Ghost Hunters: William James and the Search for Scientific Proof of Life After Death.* New York: Penguin, 2006

Brandon, Ruth. *The Spiritualists: The Passion for the Occult in the Nineteenth and Twentieth Centuries.* New York: Knopf, 1983

Burpo, Todd, with Vincent, Lynn. *Heaven is for Real: A Little Boy's Astounding Story of His Trip to Heaven and Back.* Nashville: Nelson, 2010

Carrington, Hereward. *The Problems of Psychical Research: Experiments and Theories in the Realm of the Supernormal.* New York: Dodd, Mead, 1921

Cayce, Edgar Evans. *Edgar Cayce on Atlantis.* London: Howard Baker, 1969

Chopra, Deepak. *Life After Death: The Burden of Proof.* New York: Harmony, 2006

Christopher, Milbourne. *Houdini: The Untold Story.* New York: Crowell, 1969

Christopher, Milbourne. *Search for the Soul.* New York: Crowell, 1979

Currie, Ian. *You Cannot Die: The Incredible Findings of a Century of Research on Death.* Toronto: Methuen, 1978

Davenport, Reuben Briggs. *The Death-Blow to Spiritualism: Being the True Story of the Fox Sisters, as Revealed by Authority of Margaret Fox Kane and Catherine Fox Jencken.* New York: G.W. Dillingham, 1888

Defoe, Daniel. *An Essay on the History and Reality of Apparitions.* London: Roberts, 1727

Diliberto, Gioia. "Patience Worth: Author from the Great Beyond." *Smithsonian Magazine*: September 2010

Dingwall, Eric J. "The Need for Responsibility in Parapsychology: My Sixty Years in Psychical Research." In Angoff, Allan, and Shapin, Betty (eds), *A Century of Psychical Research.* New York: Parapsychology Foundation, 1971

Dingwall, Eric J., Goldney, Kathleen M., and Hall, Trevor H. *The Haunting of Borley Rectory.* London: Duckworth, 1956

Dittrich, Luke. "The Prophet: An Investigation of Eben Alexander, Author of the Blockbuster *Proof of Heaven.*" *Esquire*, August 2013

Dunninger, Joseph. *Inside the Medium's Cabinet.* New York: Davis Kemp, 1935

Edwards, Paul. *Reincarnation: A Critical Examination.* Amherst: Prometheus, 1996

Fisher, Len. *Weighing the Soul: Scientific Discovery from the Brilliant to the Bizarre.* New York: Arcade, 2004

Flournoy, Théodore. *From India to the Planet Mars: A Study of a Case of Somnambulism with Glossolalia.* New York: Harper & Brothers, 1900; translation of *Des Indes à la Planete Mars: Étude sur un Cas de Somnambulisme avec Glossolalie* (1899)

Fodor, Nandor. *The Haunted Mind: A Psychoanalyst Looks at the Supernatural.* New York: Helix, 1959

Gauld, Alan. *The Founders of Psychical Research.* London: Kegan Paul, 1968

Glaskin, G.M. *Windows of the Mind: Discovering Your Past and Future Lives through Massage and Mental Exercise.* London: Wildwood, 1974

Glaskin, G.M. *Worlds Within: Probing the Christos Experience.* London: Wildwood, 1976

Glaskin, G.M. *A Door to Eternity: Proving the Christos Experience.* London: Wildwood, 1979

Goran, Morris. *Fact, Fraud, and Fantasy: The Occult and Pseudosciences.* South Brunswick: Barnes, 1979

Grant, Joan. *Time out of Mind.* London: Arthur Barker, 1956

Grant, Joan, and Kelsey, Denys. *Many Lifetimes.* London: Gollancz, 1967

Grant, John. *Corrupted Science*. Wisley: AAPPL, 2007

Grant, John. *Discarded Science*. Wisley: AAPPL, 2006

Grant, John. *Dreamers: Travellers in the World of Dreams*. Bath: Ashgrove, 1984

Green, Celia. *Out-of-the-Body Experiences*. London: Hamish Hamilton, 1968

Grey, Margot. *Return from Death: An Exploration of the Near-Death Experience*. London: Arkana, 1985

Gurney, E., Myers, F. W. H., and Podmore, F. *Phantasms of the Living*. London: Society for Psychological Research and Trübner, 1886

Hall, Trevor H. *The Strange Case of Edmund Gurney*. London: Duckworth, 1964

Hall, Trevor H. *The Enigma of Daniel Home: Medium or Fraud?* Buffalo: Prometheus, 1984

Hall, Trevor H. *The Spiritualists: The Story of Florence Cook and William Crookes*. London: Duckworth, 1962; reprinted as *The Medium and the Scientist*. Buffalo: Prometheus, 1984

Hare, Robert. *Experimental Investigation of the Spirit Manifestations, Demonstrating the Existence of Spirits and Their Communion with Mortals*. New York: Partridge & Brittan, 1855

Harris, Melvin. *Investigating the Unexplained*. Amherst: Prometheus, 2003; reissue of *Sorry—You've Been Duped* (1986)

Haynes, Renée. *The Society for Psychical Research 1882–1982*. London: Macdonald, 1982

Home, Daniel Dunglas. *Lights and Shadows of Spiritualism*. London: Virtue, 1877

Houdini, Harry. *Miracle Mongers and Their Methods: A Complete Exposé*. New York: Dutton, 1920

Houdini, Harry, with Eddy, C.M. Jr. (uncredited). *A Magician Among the Spirits*. New York: Harper, 1924

Huston, Peter. *Scams from the Great Beyond*. Boulder: Paladin, 1997

Iverson, Jeffrey. *More Lives Than One?: The Evidence of the Remarkable Bloxham Tapes*. London: Souvenir, 1976

Jastrow, Joseph. *Fact and Fable in Psychology*. Boston: Houghton Mifflin, 1900

Jastrow, Joseph. *Error and Eccentricity in Human Belief*. New York: Dover, 1962; retitled reissue of *Wish and Wisdom: Episodes in the Vagaries of Belief* (1935)

Jordan, David Starr. *The Stability of Truth: A Discussion of Reality as Related to Thought and Action*. New York: Holt, 1911

Jordan, David Starr. *The Higher Foolishness, with Hints as to the Care & Culture of Aristocracy, Followed by Brief Sketches on Ecclesiasticism, Science & the Unfathomed Universe*. Indianapolis: Bobbs-Merrill, 1927

Keene, M. Lamar, with Spraggett, Allen. *The Psychic Mafia*. New York: St. Martin's, 1976

Kelly, Lynne. *The Skeptic's Guide to the Paranormal*. Crows Nest: Allen & Unwin, 2004

Kübler-Ross, Elisabeth. *On Life After Death*, expanded edn. Berkeley: Celestial Arts, 2008

Kurtz, Paul (ed.). *A Skeptic's Handbook of Parapsychology*. Buffalo: Prometheus, 1985

Lodge, Oliver. *Raymond, or Life and Death: With Examples of the Evidence for Survival of Memory and Affection After Death*. London: Methuen, 1916

Lodge, Oliver. *Past Years: An Autobiography*. London: Hodder & Stoughton, 1931

Lodge, Oliver. *My Philosophy; Representing My Views on the Many Functions of the Ether of Space*. London: Benn, 1933

Long, Jeffrey, and Perry, Paul. *Evidence of the Afterlife: The Science of Near-Death Experiences*. New York: HarperCollins, 2010

MacLaine, Shirley. *Out on a Limb*. New York: Bantam Doubleday Dell, 1983

McMahon, Joanne D. S., and Lascurain, Anna M. *Shopping for Miracles: A Guide to Psychics and Psychic Powers*. Los Angeles: Roxbury Park, 1997

Malarkey, Kevin, and Malarkey, Alex. *The Boy who Came Back from Heaven: A True Story*. Carol Stream: Tyndale House, 2010

Marks, David, and Kammann, Richard. *The Psychology of the Psychic*. Buffalo: Prometheus, 1980

Mauskopf, Seymour H., and McVaugh, Michael R. *The Elusive Science: Origins of Experimental Psychical Research*. Baltimore: Johns Hopkins, 1980

Millard, Joseph. *Edgar Cayce: Mystery Man of Miracles—Who Saw Tomorrow, Today and Yesterday*. Greenwich: Gold Medal, 1967

Moody, Raymond A. *Life After Life: The Investigation of a Phenomenon—Survival of Bodily Death*. Covington: Mockingbird, 1975

Moody, Raymond A. *Reflections on Life After Life*. Covington: Mockingbird, 1977

Moody, Raymond A. *The Last Laugh: A New Philosophy of Near-Death Experiences, Apparitions, and the Paranormal*. Charlottesville: Hampton Roads, 1999

Moody, Raymond A., with Perry, Paul. *Paranormal: My Life in Pursuit of the Afterlife*. New York: HarperOne, 2012

Newton, Michael. *Journey of Souls: Case Studies of Life Between Lives.* St. Paul: Llewellyn, 1994

Newton, Michael. *Destiny of Souls: New Case Studies of Life Between Lives.* St. Paul: Llewellyn, 2001

Oppenhcim, Janet. *The Other World: Spiritualism and Psychical Research in England, 1850–1914.* Cambridge: Cambridge University Press, 1985

Osis, Karlis. *Deathbed Observations by Physicians and Nurses.* New York: Parapsychology Foundation, 1961

Osis, Karlis, and Haraldsson, Erlandur. *At the Hour of Death.* New York: Avon, 1977

Owen, Iris M., and Sparrow, Margaret. *Conjuring up Philip: An Adventure in Psychokinesis.* New York: Harper & Row, 1976

Palmer, Charles. *Spiritual Truth for the Young.* Manchester: Two Worlds Publishing Company, 1950

Pike, Bishop James. *The Other Side: An Account of My Experience with Psychic Phenomena.* Garden City: Doubleday, 1968

Playfair, Guy Lyon. *This House is Haunted: An Investigation of the Enfield Poltergeist.* London: Souvenir, 1980

Prince, Walter Franklin. *The Case of Patience Worth: A Critical Study of Certain Unusual Phenomena.* Boston: Boston Society for Psychic Research, 1927

Ramster, Peter. *In Search of Lives Past: Amazing New Evidence.* Sydney: Somerset Film and Publishing, 1992

Randi, James. *Flim-Flam!: The Truth about Unicorns, Parapsychology, and Other Delusions.* New York: Lippincott & Crowell, 1980

Randi, James. *James Randi: Psychic Investigator.* London: Boxtree, 1991

Raudive, K. *Breakthrough: An Amazing Experiment in Electronic Communication with the Dead.* Gerrard's Cross: Colin Smythe, 1971

Rawcliffe, D.H. *The Psychology of the Occult.* London: Ridgway, 1952; later Dover reprints are titled *Illusions and Delusions of the Supernatural and the Occult* and *Occult and Supernatural Phenomena*

Rawlings, Maurice. *Beyond Death's Door.* London: Sheldon, 1978

Ring, Kenneth. *Life at Death: A Scientific Investigation of the Near-Death Experience.* New York: Coward, McCann & Geoghegan, 1980

Ritchie, George G. *My Life After Dying: Becoming Alive to Universal Love.* Charlottesville: Hampton Roads, 1991; reissued as *Ordered to Return: My Life After Dying*, 1998

Ritchie, George G., and Sherrill, Elizabeth *Return from Tomorrow: A Psychiatrist Describes his Own Revealing Experience on the Other Side of Death*. Old Tappan: Revell, 1978

Roach, Mary. *Spook: Science Tackles the Afterlife*. New York: Norton, 2005; in the UK this book is titled *Six Feet Over*

Rogo, D. Scott. *On the Track of the Poltergeist*. Englewood Cliffs: Prentice-Hall, 1986

Rogo, D. Scott, and Bayless, Raymond. *Phone Calls from the Dead: The Results of a Two-Year Investigation into an Incredible Phenomenon*. Englewood Cliffs: Prentice-Hall, 1979

Roll, William, and Storey, Valerie. *Unleashed: Of Poltergeists and Murder: The Curious Story of Tina Resch*. New York: Simon & Schuster, 2004

Sabom, Michael. *Recollections of Death: A Medical Investigation*. New York: Harper & Row, 1982

Schatzman, Morton. *The Story of Ruth*. New York: Putnam, 1980

Shroder, Tom. *Old Souls: The Scientific Evidence for Past Lives*. New York: Simon & Schuster, 1999

Smith, F. LaGard. *Out on a Broken Limb*. Eugene: Harvest House, 1986

Snow, Chet B. *Mass Dreams of the Future: Do We Face an Apocalypse or a Global Spiritual Awakening? The Choice is Ours*. New York: McGraw-Hill, 1989

Spraggett, Allen, with Rauscher, William V. *Arthur Ford: The Man who Talked With the Dead*. New York: New American Library, 1973

Stearn, Jess. *Edgar Cayce: The Sleeping Prophet*. Garden City: Doubleday, 1967

Stearn, Jess. *Intimates Through Time: Edgar Cayce's Mysteries of Reincarnation*. San Francisco: Harper & Row, 1989

Stevens, William Oliver. *The Mystery of Dreams: A Book of True Dreams and Their Psychic Importance*. New York: Dodd, Mead, 1949

Stevenson, Ian. *Twenty Cases Suggestive of Reincarnation*. Charlottesville: University of Virginia Press, 1966

Stevenson, Ian. *European Cases of the Reincarnation Type*. Jefferson: McFarland, 2003

Strassman, Rick. *DMT: The Spirit Molecule: A Doctor's Revolutionary Research into the Biology of Near-Death and Mystical Experiences*. Rochester: Park Street Press, 2001

Thouless, Robert. *From Anecdote to Experiment in Psychical Research*. London: Routledge & Kegan Paul, 1972

Twining, Harry LaVerne. *The Physical Theory of the Soul: A Presentation of Psychic Phenomena from the Physical and Scientific Standpoint in Order to Form a Real Basis upon which to Build a Logical and Probable Theory of the Constitution of the Soul, and a Real Scientific Explanation of It's [sic] Phenomena*. Westgate, CA: self-published, 1915

Tyrrell, G.N.M. *Apparitions*. London: Society for Psychological Research and Duckworth, 1943

Wallace, Alfred Russel. *My Life: A Record of Events and Opinions*. London: Chapman & Hall, 1905

Watson, Lyall. *The Romeo Error: A Matter of Life and Death*. London: Hodder, 1974

Weiss, Brian L. *Many Lives, Many Masters*. New York: Fireside, 1988

Wicker, Christine. *Not in Kansas Anymore: A Curious Tale of how Magic is Transforming America*. San Francisco: HarperSanFrancisco, 2005

Williams, Muriel and Williams, Bill. *Life in the Spirit World: The Mind Does Not Die*. Victoria: Trafford, 2006

Wilson, Colin. *Poltergeist!: A Study in Destructive Haunting*. London: New English Library, 1981

Wilson, Colin. *Afterlife: An Investigation of the Evidence for Life After Death*. London: Harrap, 1985

Wilson, Ian. *The After Death Experience: The Physics of the Non-Physical*. New York: Morrow, 1987

Wiseman, Richard. *Paranormality: Why We See What Isn't There*. London: Pan Macmillan, 2011

Yost, Casper S. *Patience Worth: A Psychic Mystery*. New York: Holt, 1916

Zukav, Gary. *The Seat of the Soul*. New York: Fireside, 1989

INDEX